Basic Science and Art of Aircraft Wreckage Reconstruction

Basic Science and Art of Aircraft Wreckage Reconstruction

INTERNATIONAL®

Warrendale, Pennsylvania, USA

400 Commonwealth Drive
Warrendale, PA 15096-0001 USA
E-mail: CustomerService@sae.org
Phone: 877-606-7323 (inside USA and Canada)
724-776-4970 (outside USA)
FAX: 724-776-0790

Library of Congress Catalog Number 2019934419
http://dx.doi.org/10.4271/9780768099597

ISBN-Print 978-0-7680-9957-7
ISBN-PDF 978-0-7680-9959-1
ISBN-epub 978-0-7680-9961-4
ISBN-prc 978-0-7680-9960-7
ISBN-HTML 978-0-7680-9962-1
ISBN-Softbound 978-1-7680-9958-4

To purchase bulk quantities, please contact: SAE Customer Service

E-mail: CustomerService@sae.org
Phone: 877-606-7323 (inside USA and Canada)
724-776-4970 (outside USA)
Fax: 724-776-0790

Visit the SAE International Bookstore at books.sae.org

Chief Product Officer
Frank Menchaca

Publisher
Sherry Dickinson Nigam

Director of Acquisitions -
Monica Nogueira

Director of Content Management
Dan Plofchan

Senior Production Manager
Kelli Zilko

Production Associate
Erin Mendicino

Manufacturing Specialist
Christie Inman

dedication

In memory of …

Donald R. Knutson

Paul C. Jensen

Merlin K. Meisinger

John E. Ward

and Eddie E. Webber

contents

Acknowledgments

I want to proudly thank the following friends who contributed quality insight to the technical aspects of aircraft accident investigation along with the clarity of how this book was written:

John C. Clark, retired NTSB Director of the Office of Aviation Safety, who has shared a wealth of technical knowledge with me over the numerous years I have been an aircraft accident investigator.

Dean M. Thompson, retired Principal Safety Management System Leader at Textron Aviation, who was my supervisor and mentor when I worked at Beechcraft as an air safety investigator.

Steven J. Meyers, President and Chief Technical Officer of DVI Aviation, whom I have worked beside on numerous investigations as an aircraft accident consultant.

Jeffrey A. Pretz, Propulsion Certification Engineer for the FAA Small Airplane Directorate, who also worked as a fuel system design engineer at Beechcraft.

Charles "Duffy" Buchanan (C. Duffy Buchanan P.C.), an aviation defense attorney since 1982 whom I have worked with on several aircraft accident investigations.

Manuel Raefsky, Metallurgical Engineering consultant, who also worked for Westinghouse and Boeing, and has been involved with aviation safety and accident investigations since 1959.

Gerald D. Staab, retired Air Safety Investigator and Manager at Beechcraft, who teaches at the Transportation Safety Institute, and was another one of my mentors when I worked at Beechcraft as an air safety investigator.

Also, I want to thank all the individuals, government agencies, and other organizations dedicated to promoting safety (who are referenced throughout this book) for sharing their investigation documents, photos, and research data, which have helped me better present and explain the various subject matters.

Introduction

A renowned aerodynamics pioneer, Richard T. Whitcomb,* made it possible for military and civil aircraft to fly faster, farther, and with less fuel. He said, "A lot of very intelligent people are willing to adapt, but only to a certain extent. If a human mind can figure out a better way to do something, let's do it." Also, colleagues of Mr. Whitcomb said that he could "see air" and relied more upon wind tunnel experiments than mathematical solutions. I appreciate Mr. Whitcomb's ability to visualize complex circumstances and his drive to find practical solutions using a hands-on approach. Similarly, I hope that this book promotes improving the way investigators examine aircraft wreckage and visualize how the crash occurred, especially when stored flight or systems data (e.g., flight data recorders [FDRs or "the black box"] or other instrumentation with non-volatile memory) is unavailable or cannot be downloaded due to crash damage.

There has been only a handful of publications that comprehensively address "how to" evaluate aircraft wreckage such as within International Civil Aviation Organization (ICAO)† Document 9756, Aircraft Accident and Incident Investigation, Part III. This book is intended to *enhance* existing aircraft accident investigation (AAI) publications by providing practical insights on wreckage reconstruction, drawing from my 37+ years of overall aviation experience that includes AAI (airplane and rotor-wing), mechanical engineering, airframe and powerplant maintenance, piloting, and developing/instructing AAI courses.

---

* Mr. Whitcomb (1921–2009) was an engineer for the National Aeronautics and Space Administration (NASA) who conceived and developed three revolutionary aerodynamic concepts – the "area rule," "supercritical wing," and "winglets."

† ICAO is a United Nations specialized agency established in 1944 that works with 192 member states (countries) and industry groups to develop international civil aviation standards and recommended practices and policies in support of safe, efficient, secure, economically sustainable, and environmentally responsible civil aviation operations.

This book will discuss [i] some pertinent aspects on investigating crashes as well as who investigates them and why, [ii] healthy attitudes and philosophies required to properly perform investigations, [iii] what to expect from the on-scene part of the investigation, and primarily [iv] fundamental approaches to common types of wreckage reconstruction.

Many aircraft crash and failure scenarios will be presented in a "generic" form but relate to actual investigations. Every possible scenario cannot be covered; however, the various examples presented provide investigative concepts that can be applied to other aircraft/systems/components that failed or operated under similar circumstances. Investigation methods and techniques presented are also intended to be basic in nature. For example, the chapter on fire and explosions includes a general approach to evaluate in-flight versus post-impact fires and what type of explosion occurred. However, there are no heat transfer calculations or detailed discussions on how to determine chemical composition of a material (albeit useful information). Similarly, this book provides ways to evaluate structural failures and estimate engine power conditions, but does not provide engineering strength of materials calculations or comprehensive powerplant teardown inspection procedures.

Basic Science and Art of Aircraft Wreckage Reconstruction is intended for the AAI practitioner, student, or someone who is simply curious about how aircraft wreckage is reconstructed. Subject references are provided in the various chapters; however, the reader needs to possess knowledge or be willing to learn about aircraft structures, airframe and powerplant systems, aircraft flight control principles, aviation operations, and basic physics and math (algebra, geometry, and trigonometry). Also, a certain amount of research and looking up definitions will most likely be necessary due to the technical nature and diversity of the subjects covered.

A great deal of thought and care went into not sensationalizing this book's contents. In summary, it should prove as a useful investigative resource whether you are an apprentice or veteran with a government aviation agency (NTSB, AAIB, FAA, etc.),* an aircraft/engine/component manufacturer, military branch, insurance company, law enforcement agency, or law firm.

* National Transportation Safety Board (NTSB) has jurisdiction in the United States with responsibilities to [i] make rules and regulations governing accident notification and reporting of accidents involving civil aircraft; [ii] investigate the accidents, report the findings, and try to determine the probable cause; [iii] make safety recommendations to the FAA: [iv] provide reports that the NTSB deems in the public interest; and [v] conduct special studies/investigations to improve safety and reduce accidents.

- Air Accidents Investigation Branch (AAIB) is similar to the NTSB and has jurisdiction in the United Kingdom.
- When an accident occurs in the United States, the Federal Aviation Administration (FAA) is responsible for investigating [i] violations of the Federal Aviation Regulations (FARs); [ii] performance of FAA facilities or functions involved; [iii] airworthiness of FAA-certificated aircraft involved; [iv] competency of FAA-certificated airman, businesses or airports involved; [v] FARs are/were adequate; [vi] airport certification standards or operations involved; [vii] air-carrier or airport security standards or operations involved; [viii] airmen medical qualifications involved; and [iv] performance of non-FAA owned and operated ATC facilities or navigational aids.

CHAPTER 1

Perspectives on Investigations and Investigators

CHAPTER 1

Basic Anatomy of an Accident and Its Reconstruction

An accident* (or mishap, as used in military circles) always involves the person (human), the machine, and the environment (Figure 1.1). Just like starting a fire requires a source of heat, fuel, and oxygen – if any one of these ingredients is removed, then nothing is going to burn. Similarly, an accident/mishap involves a chain of events that lead to an ultimate catastrophe. Figure 1.2 shows how a series of realistic circumstances can build on themselves and lead to a crash – if the pilot broke any part of the chain by following the proper procedures or had maintained his/her proficiency, then the accident most likely would not have even occurred. Accident reconstruction is fundamentally based on evaluating all available human-machine-environment circumstances along with establishing the sequence or timeline of events leading up to the crash or failure.

The definition of "reconstruction" (as it relates to investigating accidents) is *to establish or assemble again* or *to build up mentally*.† This means that *wreckage* reconstruction is directly related to "inspecting" or "examining" wreckage, and each of these verbs apply to understanding what the aircraft's condition and circumstances were before, during, and after the accident.

---

* Code of Federal Regulations (CFR), Title 49, section 830.2, defines an "accident" as an occurrence associated with the operation of an aircraft which takes place between the time any person boards the aircraft with the intention of flight and all such persons have disembarked, and in which any person (either inside or outside the aircraft) suffers death or serious injury, or in which the aircraft receives substantial damage.

† Definition from the Merriam-Webster dictionary.

FIGURE 1.1 Ingredients for an accident.

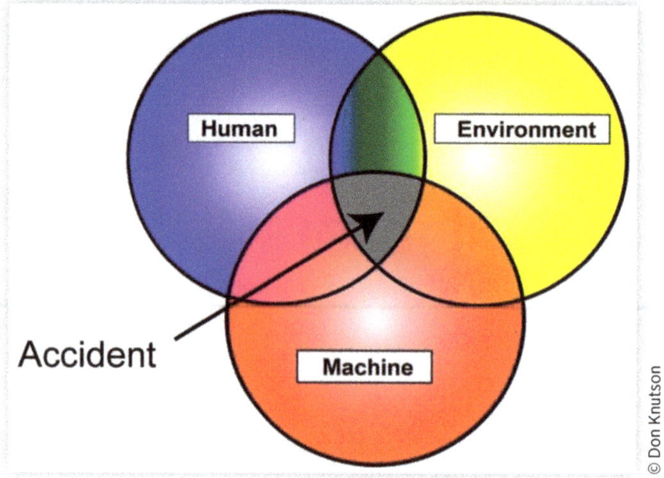

FIGURE 1.2 A chain of events needs to occur before an accident occurs.*

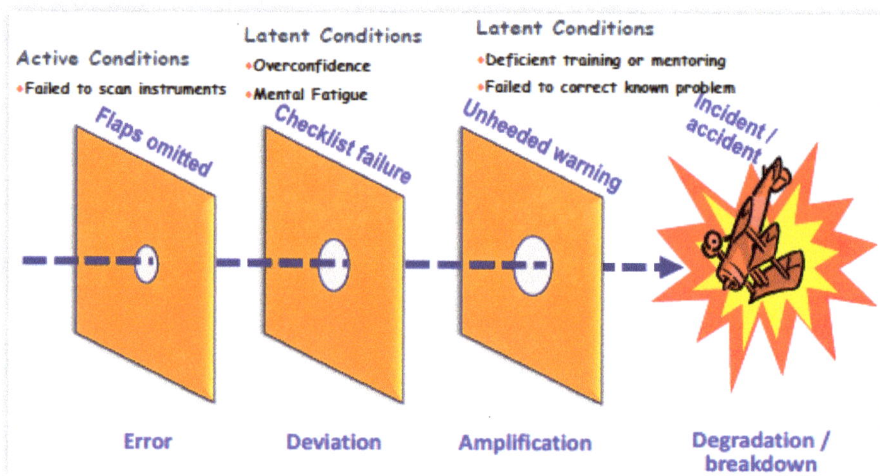

Wreckage reconstruction can mistakenly be associated with *just* the "machine" ingredient for the accident – the hands-on or "kickin tin" aspects of examining parts. In fact, the "human" and "environment" circumstances help determine the direction and depth of the wreckage inspection. For instance, crashworthiness analysis (discussed in Chapter 5) involves more than just evaluating the airframe damage and resultant impact (G) loads on the occupants. Assume that a crash investigation involved a pilot that had to ditch an aircraft (in the water). Was an emergency briefing card onboard and accessible to the passengers in the aircraft? Was the aircraft properly equipped with life vests or a raft, and were they used? Were the cabin doors opened before water impact to allow a quicker exit? If the doors were closed, did airframe impact deformation cause them to be difficult to open? These practical questions can be answered via the aircraft wreckage exam and are meaningfully associated with evaluating the "human" (emergency procedures) and "environment" (cockpit/cabin) ingredients for the accident.

* Diagram adapted from the "Swiss cheese model" theory presented by James T. Reason, PhD, professor of psychology at The University of Manchester. Also, reference "Revisiting the Swiss Cheese Model of Accidents," Eurocontrol Experimental Center, Note No. 13/06, issued October 2006.

Why Are Aircraft Accidents Investigated?

The most obvious reasons are to establish who/what caused the accident and use the lessons learned to try to prevent or reduce accidents from occurring in the future. Also, family or friends of the deceased or injured want to find closure from the tragedy, and people or organizations want to recover or avoid losses/damages (which usually involves a lawsuit). These reasons bring up an interesting thought about who investigates accidents and how these investigations are conducted. For example, a civilian accident investigation in the United States involves the NTSB, FAA, and companies that manufacture, operate, maintain, or support the subject aircraft (who are given NTSB party status* to provide technical assistance). The NTSB has no authority to regulate, fund, or be directly involved in the operation of any mode of transportation and is supposed to conduct investigations and make recommendations from an objective viewpoint. On the other hand, the FAA has the responsibility to determine whether any regulations were violated, that is, identify and consider penalizing those involved in the accident. Also, the companies or organizations represented by the nongovernment investigators are typically faced with potential product liability.

Regardless of whom the investigators represent, they should maintain their ethical responsibility of finding the truth. For a variety of reasons, organizations and individuals that become involved with litigation can face strong temptation to "shoot the arrow and paint the bull's-eye around wherever it lands," that is, investigators can have a predetermined mission to develop or fabricate a story based on whatever is found during the investigation (no matter how unsubstantiated) to establish fault or innocence.

The Human Factors Aspect

Causes or contributing factors of accidents/incidents involving human beings always need to be factored into the equation.† The clear majority of aviation accidents are at least partially attributable to human error. Quantifying physical evidence such as material properties and the laws of physics is straightforward – strength, melting point, gravity, etc. stay the same no matter what sequence of events or human interactions occur. Conversely, analyzing how/when/why a person acts or reacts before, during, and after an event can be very difficult because it depends on so many variables such as overall life experience, health, personality, and level of training. Human limitations (e.g., recognition and reaction times, vision clarity, motion perception without visual cues, and g-load tolerance) are predicable from a minimum or maximum amount obtained from researched and tested performance capabilities. In many cases, without devices that specifically record an event, we are left with evaluating what a person did or did not do from a "more likely than not" standpoint based on empirical research and past accident data with similar circumstances.

* Rules for NTSB investigation party status are defined in 49 CFR § 831.11.

† *A Human Error Approach to Aviation Accident Analysis*, Wiegmann and Shappell, 2003 – this reference provides the Human Factors Analysis and Classification System, which is intended to systematically break down the various aspects of why someone does something under certain conditions or events.

Requirements to Become an Investigator

Effective "formal" education is available to train aircraft accident investigators (both basic and advanced),* but no certification exists that specifically qualifies someone to investigate aircraft accidents (at least at the time of writing this book). The legal system requires "experts" to have specific experience and/or training commensurate to the opinions they will offer in testimony. Also, there is a significant percentage of aircraft accident investigation (AAI) "experts" who derive most (if not all) of their investigation experience from getting hired on legal cases.

Investigators typically start out with a specific background in aviation such as a pilot, mechanic, and/or engineer. ISASI has guidelines for investigator training and education,† which lists the following appropriate areas of knowledge:

- Mathematics and science
- Verbal, written, and technical report writing/communication skills
- Cognitive analytical skills
- Computer science
- Engineering
- Management
- Psychology

From "the world according to Don," the minimum requirements to perform AAI should be [i] reasonable hands-on aviation experience in engineering, maintenance, manufacturing, and/or piloting; [ii] a passion for investigating and problem-solving, even when the work becomes tedious; [iii] the ability to effectively present and report findings, which logically support conclusions and can provide useful information or recommendations to the decision-makers; and [iv] self-motivation to take/attend continuing education classes and seminars, which regularly furthers one's knowledge in air safety and investigative methods and techniques. As you can see, this definition emphasizes one's overall character and attitude as well as their aviation experience.

Obtaining at least a private pilot's certificate is recommended because it requires you to basically understand the piloting eligibility requirements, airmen privileges and limitations, aircraft certificates and documents needed to operate an aircraft, weather, aircraft aerodynamic performance and limitations, aircraft systems and operations, flight planning and navigation, aeromedical factors, and aeronautical decision-making. However, having a pilot's license is not essential to performing an AAI. A good example of this is an aircraft mechanic who needs to understand the concept of how the aircraft operates but is not required to have piloting skills to troubleshoot and fix the airplane. Similarly, an aeronautical engineer can understand how to design and manufacture an aircraft without specifically being certified and rated to fly it. As a matter of fact, many

* Civilian organizations that have established AAI training programs include Cranfield University (Cranfield, United Kingdom), University of Southern California (Los Angeles, CA), Southern California Safety Institute (Long Beach, CA), Embry-Riddle Aeronautical University (Daytona, Beach, FL, and Prescott, AZ), Transportation Safety Institute (Oklahoma City, OK), and NTSB Training Center (Ashburn, VA).

† www.isasi.org, ISASI Guidelines for Investigator Training and Education, First Edition – August 1999.

great pioneers of aviation safety and innovation were not pilots, such as Jerome F. Lederer,* Charles E. Taylor,† and Richard T. Whitcomb.

Attending both a basic and advanced AAI course is highly recommended for new investigators; however, the best overall education comes from "hands-on" training along with working closely with experienced investigators who want to mentor others in both the science and art of AAI. Several educational institutions that offer AAI courses include "laboratory" training, where the students have an opportunity to "kick tin" in a simulated accident environment. As a teacher and mentor, I think this approach is effective and have used this concept in courses that I have developed and taught. Another excellent AAI lab exercise (although impractical to provide) would be to receive a surprise phone call at 2:00 am on a Saturday morning that notifies you that an aircraft crashed in remote mountainous terrain; and you need to immediately make airline travel arrangements (that typically requires at least 6-8 hours travel time domestically, and much more if you need to fly internationally). When arriving at the airport, you travel at least another hour or two to meet your investigation team, and then be transported by helicopter to the accident site. The accident scene would be located on a deep snow pack at an approximate 13,000-ft elevation, with temperatures close to or below zero degrees Fahrenheit. Due to limited daylight hours, you and your investigation team would only have 3 to 4 h before the helicopter needed to pick you up. Oh … by the way, the helicopter needs to land on top of the ridge line at about 400 yards from the accident site due to no available safe landing zones on the snowpack. Therefore, you need at least 30 min each way to traverse the waist-deep snow while carrying investigation gear to and from the accident scene. You also would need to be aware of the physical limitations of breathing at high altitudes, which can cause dizziness, fatigue, headaches, and nausea. With all this said, now you can start concentrating on documenting the wreckage and accident site. If not remote mountainous terrain, it can be a hot, humid, swamp in the jungle dealing with the threat of poisonous snakes and flies or mosquitoes constantly biting you, or in the middle of a scorching desert where you cannot seem to drink enough water, fine granular sand blowing into every orifice of your body, and most anything you touch exposed to the sun burns your fingers. In summary, field investigators who are required to be on-scene after the accident occurs need to accept working long hours over several consecutive days in unfamiliar places while enduring varying physical and emotional challenges.

Generalists and Specialists

Investigators with diverse aviation experience in engineering, maintenance, and piloting typically make effective aircraft accident reconstructionists or generalists because of their overall understanding of aircraft design and operations. The more we learn, the more we realize we don't know; therefore, generalists need to continually develop their

* Mr. Lederer (1902–2004) was an American engineer who is known as "Mr. Aviation Safety." He performed groundbreaking safety studies for the Air Mail Service, the Civil Aeronautics Board's Safety Bureau, and the National Advisory Committee for Aeronautics. He's credited with introducing blinking anti-collision lights and flight data recorders (FDRs). He also participated in the group that organized the FAA and modernized the air traffic control system.

† Mr. Taylor (1868–1956) was an American aircraft mechanic who built the first engine used by the Wright brothers and was a vital contributor in the building and maintaining of early Wright engines and airplanes.

understanding of how to properly utilize and support a multitude of specialists (when necessary) who have the following areas of expertise:

- Air traffic control (ATC) and radar data
- Airport operation and design
- Avionics and instrumentation (including the extraction of stored data from non-volatile memory chips)
- Biomechanics or kinesiology (body movements, injury-producing mechanisms, etc.)
- Crashworthiness (protection of aircraft occupants during and after a crash)
- Fire and explosion (heat transfer and chemical analysis, explosive devices, etc.)
- Flight data recorder (FDR) and cockpit voice recorder (CVR) readouts and analysis
- Human factors (machine-person-environment interface issues)
- Maintenance (airframe and powerplant)
- Manufacturing and design (how the aircraft was built and what are its limitations)
- Materials (metals, composites, plastics, etc.)
- Meteorology (weather data and studies)
- Pathology and toxicology (autopsy and drug testing)
- Piloting (flight test, instruction, or general operation)
- Simulator and animation (visualizing and presentation of events)
- Sound spectrum analysis (evaluating background noise frequencies from radio transmission recordings and CVR)
- Systems engineering (electrical, flight control, fuel/propulsion, structural, etc.)
- Test and system modeling (proving a theory or verifying a finding)
- Tribology (lubrication, friction and wear)

"Heavy Iron" Versus Most General Aviation Investigators

The number of general aviation (GA)* aircraft in service along with their accident statistics is considerably greater than the number of transport category aircraft.† Furthermore, when these "heavy iron" aircraft crash, the investigators commonly have the luxury of

* GA is defined by ICAO as "all civil aviation operations other than scheduled air services and non-scheduled air transport operations for remuneration or hire."

† Jets with ten or more seats or a maximum takeoff weight (MTOW) greater than 12,500 pounds, propeller-driven airplanes with greater than 19 seats or a MTOW greater than 19,000 lb, or helicopters with a MTOW greater than 7,000 lb.

additional resources such as a "Go-Team,"* along with a FDR and CVR to aid them in their investigations. FDRs and CVRs typically provide a great deal of immediate answers to what happened to the aircraft and pilots before, during, and, in many cases, after the crash or failures occur. Most GA investigators do not have a Go-Team or FDRs and CVRs at their disposal, which appreciably increases the challenge of investigating an accident or mishap, especially when considering that GA consistently has the vast majority of accidents. The science of investigating accidents does not change, but GA investigations present challenges that a perspicacious investigator must deal with. This situation is analogous to a person who shops at the grocery store to buy food versus someone who primary lives off the land hunting, fishing, and farming for what they eat. This does not necessarily mean that the person who goes to the grocery store does not understand how to hunt, fish, and farm. However, the person who is forced to live off the land inherently develops and maintains a greater scope of various skill sets that are not required when just picking food off a shelf and buying it.

When Consultants/Experts Become Involved

Insurance companies and law firms will typically hire consulting experts† or testifying expert witnesses (consultants/experts) when people or organizations want to recover losses/damages (due to the death of loved ones, loss of property, lost earnings, etc.). This usually occurs after the government agencies conclude their field investigation and release the wreckage. Consultants/experts have at times discovered profound information that proves findings reported by the government investigation agencies to be incomplete or inaccurate. Among the reasons for this is that litigation often results in more

* NTSB defines an aviation "Go-Team" as a group of investigators under the leadership of an investigator-in-charge (IIC), with the following noted specialties and responsibilities:

Operations – the history of the accident flight and crewmembers' duties for as many days prior to the crash as appears relevant

Structures – documentation of the airframe wreckage and the accident scene, including calculation of impact angles to help determine the aircraft's pre-impact course and attitude

Powerplants – examination of engines (and propellers) and engine accessories

Systems – study of components of the aircraft's hydraulic, electrical, pneumatic, and associated systems, together with instruments and elements of the flight control system

Air traffic control (ATC) – reconstruction of the air traffic services given the plane, including acquisition of ATC radar data and transcripts of controller-pilot radio transmissions

Weather – gathering of all pertinent weather data from the National Weather Service, and sometimes from local TV stations, for a broad area around the accident scene

Human performance – study of crew performance and all before-the-accident factors that might be involved in human error, including fatigue, medication, alcohol, drugs, medical histories, training, workload, equipment design, and work environment

Survival factors – documentation of impact forces and injuries, evacuation, community emergency planning, and all crash-fire-rescue efforts

† "Consulting experts" are typically used to assist with the litigation but are not disclosed nor will any of their work product be discoverable. They are intended to be non-testifying advisors that help attorneys better understand factual evidence, such as industry-standard practices or how a system is designed, and can assist with vetting and hiring testifying expert witnesses, preparing for depositions of the opposing side's experts, and formulating various litigation strategies.

money and expertise being directed to an investigation, which theoretically should improve the chances that more overall details about the accident will be discovered. Another advantage that consultants/experts usually have over field investigators is that they are essentially investigating the investigation – in other words, consultants/experts can typically look at issues with more hindsight and time to evaluate pertinent safety and liability issues that do not always surface during the initial field investigation.

When litigation is involved, the consultant/expert in many cases has more "credentials" than the average field investigator, such as a master's or doctorate (PhD) degree with numerous published papers. It is common for aviation consultants/experts (in metallurgy, meteorology, pathology, biomechanics, etc.) to also be involved with failures and accidents that are not related to aviation. Having "Dr." at the beginning of your name or other additional professional accomplishments is admirable and can add credibility when testifying in a deposition or trial, but keep in mind that these credentials do not necessarily improve one's ability to investigate accidents.

Space Is the Next Frontier

Commercial space travel will most likely be the next significant challenge for aerospace safety. The next generation of investigators will need to possess a strong technical knowledge of both astronautics and aeronautics. Investigators will also benefit from the ever-improving data recording and flight-tracking capabilities. These technologies are already evident with the continual improvements in microchip data storage in many types of aircraft instruments, and satellite tracking replacing ground-based radar systems.

Some of the biggest challenges will include the growing pains of developing consistently reliable and safe spacecraft that operate outside of the earth's atmosphere and regularly encounter very severe conditions (e.g., temperature swings between -200°F and 250°F, and ionizing [cancer-causing] radiation), while still requiring all the present-day systems and conditions as aircraft operating within earth's atmosphere. Figuring out design, manufacturing, and operational safety requirements necessary for more cutting-edge and complex systems (e.g., environmental, propulsion, and navigation) will probably become similar to the earlier years of aircraft development where numerous lessons needed to be learned before achieving the level of overall air safety that we enjoy today. This will inherently require more emphasis on safety management systems (SMS).*

* Refer to FAA Order 8000.369B, Safety Management System.

CHAPTER 2

Back to Basics

AAI reports with inaccurate and incomplete findings litter the "field" investigation databases, which affects our ability to rely on accurate empirical knowledge to compare past investigation findings with current ones, or to conduct meaningful safety studies. Also, the overwhelming majority of investigations are related to general aviation, which are primarily "field" investigations. Working as an air safety field investigator and accident and wreckage reconstruction consultant/expert has significantly helped me realize how inaccurate and/or incomplete investigations occur (even when investigators are honestly looking for what happened with no preconceived agenda). This chapter discusses some practical and healthy ways to approach investigations and presents a couple of actual field investigation case studies that resulted in incomplete and inaccurate findings.

Physical Evidence Is King

Even though investigators become savvier and more confident based upon their experience, we must guard against becoming overconfident or complacent. For example, a fireman with 30 years of experience might investigate a house that burned down, and the available evidence shows that the fire started somewhere in the basement. If he determines that the furnace was the origin of the fire based upon the fact that he has seen this happen hundreds of times during his career, but no evidence exists to support his conclusion, then he has simply used flawed analysis. Investigators must base their conclusions on facts and scientific principles, not just experience (refer to Figure 2.1) or "gut feelings."

Accepting witness statements from people who saw, heard, or experienced an event as factual evidence is another way investigators can lead themselves down the wrong path. I have provided student training on taking witness statements where I alerted them that they

FIGURE 2.1 An example of why we should not assume just our experience is enough to make conclusions.

© Don Knutson

were going to witness an accident on a video. Immediately after seeing the accident, I would turn off the video screen and collectively ask the class what they saw, which in many cases turned out to be no better than 50% accurate. This clearly demonstrates that our memories are typically inefficient and have difficulty with paying attention to every detail. Human factors studies have also supported this limitation, especially, after prolonged time or dealing with trauma. Therefore, a person's recollection should *only* be used as a tool for exploring *possibilities* as to whether or not the event occurred. Figure 2.2 provides a good example of why witness statements should not be used as factual evidence.

Close Calls

High visibility and liability usually drive the depth of an investigation. For instance, typical human nature is to "relax" when nothing gets destroyed and nobody gets hurt or sued. Think about how "minor" incidents that turned into close calls (e.g., a near mid-air collision, an airplane taking off that barely cleared trees at the end of a runway, and a mechanical failure that forced an uneventful emergency landing) could have just as easily turned into a catastrophic accident. These incidents‡ turn into empirical knowledge that can aid accident investigations (providing the information is complete and accurate). This concept also directly relates to the fundamental aspects of SMS as shown on Figure 2.3. Knowing what *could* have happened as well as what *really did* happen clearly helps an organization better understand what can potentially happen in the future and take measures to reduce or prevent the situation from occurring again.

FIGURE 2.2 Summation of eyewitness statements showing clear differences in what they saw.*

Excerpt from the NTSB investigation into the crash of American Airlines Flight 587:

"Witness Interviews
The Witness Group has received 349 accounts from eyewitnesses, either through direct interviews or through written statements. An initial summary of those statements follows:

- 52% specifically reported seeing a fire while the plane was in the air, with the fuselage being the most often cited location (22%). Other areas cited as a fire location were the left engine, the right engine or an unspecified engine, and the left wing, the right wing or an unspecified wing.
- 8% specifically reported seeing an explosion.
- 20% specifically reported seeing no fire at all.
- 22% reported observing smoke; 20% reported no smoke.
- 18% reported observing the airplane in a right turn; another 18% reported observing the airplane in a left turn.
- 13% observed the airplane "wobbling," dipping" or in "side to side" motion.
- 74% observed the airplane descend.
- 57% reported seeing "something" separate from the airplane; 13% reported observing the right wing, left wing or an undefined wing separate; 9% specifically reported observing no parts separate."

© National Transportation Safety Board

* NTSB/AAR-04/04, In-Flight Separation of Vertical Stabilizer, American Airlines Flight 587, Airbus Industrie A300-605R, N1405 Belle Harbor, New York, November 12, 2001.

† https://www.faa.gov/about/initiatives/sms/explained/basis/

‡ CFR, Title 49, section 830.2, defines an "incident" as an occurrence other than an accident that affects or could affect the safety of operations.

FIGURE 2.3 SMS concept where organizations incorporate reactive, proactive, and predictive thinking.*

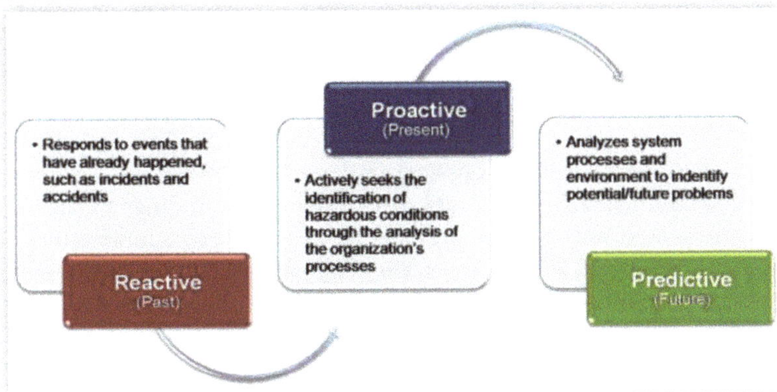

© FAA

One Piece at a Time

We should not start forming any conclusions until the facts are completely documented and evaluated. When looking at something like this,

© Don Knutson

when have you heard someone ask, "How do you take all those broken pieces and understand what happened?" The answer is, "One piece at a time." This answer seems too simplistic, but it is exactly how we need to start approaching the challenge of reconstructing the wreckage. A disciplined approach is required because our experience, knowledge, and ego can lead us astray and prompt us to cut corners or jump to conclusions. Sometimes we get away with this, but we are not exercising quality control no matter how we would like to justify it. One way to illustrate this basic investigative concept is to solve the following cryptogram (the answer is at the end of this chapter):

FG VUZRQ VTPJE EDHMFMJ ZFH RHZECJE, ACJX KCS FEX'A ACJ KCTUJ ZFHOUZXJ YZNJ TDA TG ACZA EADGG?

Z = **A**, **V** = **B**, **G** = **F**, **E** = **S**, **A** = **T**, **M** = **V**, **K** = **W**

Tools needed to solve this puzzle are knowledge of the English language (reading, writing, and spelling), similar to understanding engineering and sciences. Also, knowledge of the English-speaking culture is required (slangs, humor, etc.), similar to understanding the aviation industry. We should logically start by replacing all the given letters, similar to finding initial pieces of evidence. Next, evaluate the small words to determine

FIGURE 2.4 Scientific method flow chart.*

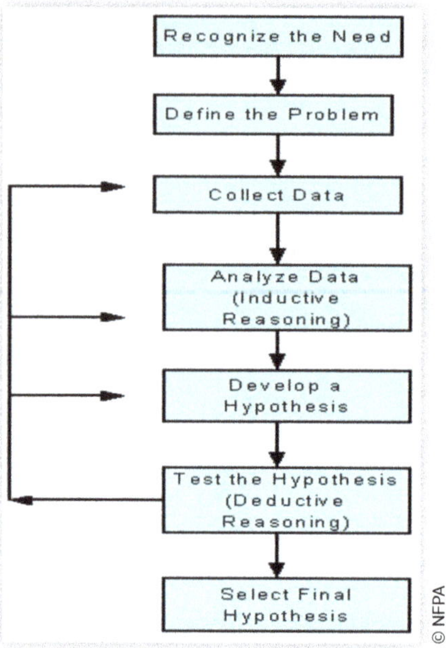

their vowels. Note that during this process, we start by concentrating on the words in the sentence and not the entire sentence. Likewise, we need to focus on pieces of wreckage when analyzing how the overall aircraft system or structure failed. After we make some educated guesses (iterations) on which letters can work, we start to string two or more words together. Then, we further iterate the process until we see that the sentence makes sense. Likewise, as we accumulate our wreckage findings piece by piece, we theorize realistic possibilities based on the available evidence. Then, we compare relationships between two or more findings and iterate through logic and tests to find out whether or not we can connect the dots. Our ultimate goal is to establish the sequence of events and causes leading to and including the accident.

The Scientific Method

A few examples will be presented in this chapter that show how inaccurate or incomplete investigation findings result from *selective gathering of evidence*, instead of truly evaluating and piecing together all the findings. To minimize the influence of any bias or prejudice in the investigation, we should evaluate hypotheses or theories through accurate, reliable, consistent, and nonarbitrary representation of the investigative findings. The flow chart depicted in Figure 2.4 concisely shows this basic process; however, effectiveness of this method is only as good as the investigative resources and completeness of information gathered. If we do not have the right expertise or our information lacks enough facts to connect the dots (i.e., supporting evidence), then this method turns into "junk in, junk out."

Nine-Box Matrix

Established checklists (e.g., ICAO Manual of Aircraft Accident and Incident Investigation, Part II, Procedures and Checklists; and Air Accidents Investigation Branch [AAIB, from the United Kingdom] Procedure Manual of Accident/Incident Investigation) or accident report formats help remind/guide us to consider important pieces of information to gather during the investigation. The nine-box matrix (Figure 2.5) is a technique that encourages the investigator to dig farther into the details by breaking down the possibilities surrounding the facts and events that occurred before, during, and after the accident. This process becomes even more important when we may not get another chance to document and evaluate specific evidence after the on-scene investigation (e.g., diminished witness recollection over time, lost or expired data, unable to conduct further autopsy exams or toxicological tests, and the aircraft wreckage being disassembled at and recovered from the crash site).

Each of the nine segments typically starts out with listing relatively general questions (or specific information to try to obtain) and evolves into a more detailed "to-do list." This concept can be used during any phase of the investigation.

* Obtained from National Fire Protection Agency (NFPA) 921 – Guide for Fire and Explosion Investigations, ISBN 9781455916023.

FIGURE 2.5 Nine-box matrix.

	Person (pilots, operators, passengers, witnesses, mechanics, etc.)	Machine (airframe, powerplant, avionics, ground support, etc.)	Environment (weather, flight path, accident scene, work culture, etc.)
Before			
During			
After			

Questions or information to obtain concerning the "Person" involves the pilots along with the passengers, air traffic controllers, mechanics, flight attendants, witnesses, etc.; "Machine" involves the aircraft systems and components along with communication/navigation aids, ground support and maintenance equipment, etc.; and "Environment" involves the weather along with airspace/flight path, airport operations, company policies/procedures, safety culture, workspace conditions, etc. These questions are analogous to pixels in a photo – the more pixels used (questions answered or information found), the greater the image resolution becomes (events before, during, and after the accident).

General questions or information to obtain regarding the *person[s]* include the following:

Mental aspects:

- Knowledge/experience?
- Training/education?
- Attitude/emotional state?

Physical aspects:

- Size (e.g., height, weight, and inseam)?
- Gender?
- Age?
- Senses affected (e.g., vision, hearing, and balance/inner ear)?
- Disabilities?

Physiological aspects:

- Health and physical conditioning?
- Lifestyle/habits?
- Alertness/fatigue?
- Chemical dependency?

Psychosocial aspects:

- Interpersonal relations?
- Ability to communicate?
- Leadership abilities?

General questions or information to obtain regarding the *environment[s]* include the following:

Physical aspects:

- Weather conditions?
- Airspace/flight path?
- Workspace conditions (e.g., in maintenance shop, in cockpit, or on service ramp)?
- Safety issues or emergency situations?

Organizational aspects:

- Job tasks?
- Supervision required?
- Morale?
- Company policies/rules?
- Safety or emergency procedures?

Typical/general questions or information to obtain regarding the *machine[s]* (remember that this involves more than just the aircraft and its components):

- Properly certified/calibrated?
- Maintenance records - properly maintained?
- Airworthiness Directives (ADs) (mandatory compliance)?
- Service Bulletins and Instructions (recommended by manufacturer)?
- Maintenance manuals and illustrated parts catalog?
- Pilot's Operating Handbook (POH) or operations manual?
- Appropriate instructions to operate?
- Flight plans and fueling records?
- Modifications or changes to original design?
- Preexisting malfunctions or missing parts?

After the general questions are listed, we look to deepen their scope. For instance, more detailed questions related to the maintenance shop environment could be as follows:

- Appropriate/up-to-date maintenance data on hand and accessible?
- Checklist used during the maintenance procedures?
- Mechanic having adequate hands-on experience to perform the maintenance?
- Shop properly organized and equipped?
- At least two sets of eyes inspecting and documenting that the work was accomplished?
- Records clearly note repairs/inspections performed and what was found before and during work accomplished?
- How often are the precision measurement tools calibrated (e.g., torque wrenches, micrometers, weight and balance scales, or pitot-static tester)?
- Maintenance personnel communications and correspondence with the aircraft pilot/owner documented (e.g., customer aware of pertinent Service Bulletins not complied with or unairworthy conditions)?

Consider Even the Absurd

Investigators need to remain open-minded to all possibilities. The following two examples of actual accident circumstances give credence to this notion.

First, let's visualize a twin-engine propeller-driven airplane flying level at cruise speed while crashing into gradually rising and densely forested terrain similar to Figure 2.6. The main wreckage comes to rest ~500 ft from the initial tree impact and sustains a post-impact fire. One of the propellers (Figure 2.7) had separated from its engine and was found in proximity to the main wreckage. Notice how the three-bladed aluminum propeller shows pronounced aft curling on *only* one blade tip, which is consistent with high engine power and propeller RPM. Also, notice that the other two blades exhibit no missing tips, no appreciable twisting, no leading edge gouging, and shallow bending in opposite directions - this damage in itself is consistent with low or no engine power and propeller RPM (Chapter 4 will specifically cover propeller blade impact damage). Think about the odds of just one blade cutting through the trees in a dense forest. In addition, if just the curled part of the blade had somehow been consumed by the post-impact fire, think about how this could create a false theory of an engine power loss occurring before terrain impact.

FIGURE 2.6 Densely forested terrain similar to where the propeller shown on Figure 2.7 had chopped through.

FIGURE 2.7 Views of a three-bladed propeller assembly with one blade curled.

A second example is an airplane flying a long cross-country during the winter with four persons on board. While cruising at 10,000 ft MSL, the pilot lost consciousness (believed to be from carbon monoxide poisoning due to circumferential cracks found on the burner can of the gas-fueled cabin heater assembly). Unknowingly, the pilot lost control of the airplane and allowed it to enter a high-speed descent. Subsequently, the pilot somehow regained his awareness and made an abrupt pull-up maneuver that resulted in an in-flight breakup of both wings and the stabilator (pitch control surface). Wreckage was scattered approximately two miles (Figure 2.8). The remaining fuselage cockpit and cabin crashed in a remote forested area, and somehow impacted trees in a way that buffered the fall enough to amazingly keep two of the occupants alive.

The above two examples show how important it is to be thorough during the wreckage examination as well as expect the unexpected.

FIGURE 2.8 Two-mile wreckage scatter from an in-flight breakup.

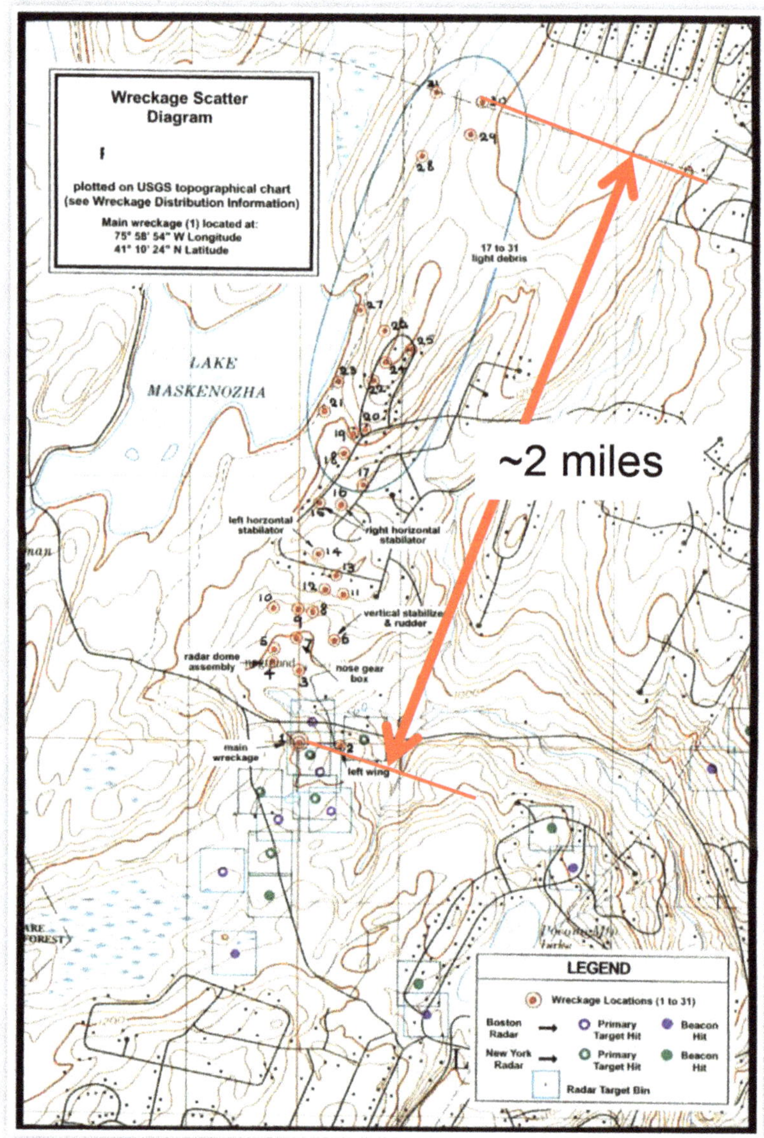

Consequences of Incomplete and Inaccurate Findings

Let's look at a few actual AAI case studies that demonstrate the importance of using sound investigation practices. The first case study involves an accident investigation with *incomplete* findings that could have led (or may lead) to unnecessary injuries in future accidents (providing no steps were [or are] taken to make improvements). The second case study shows how *inaccurate* findings led to unnecessary mandated aircraft modifications and costs that unfairly affected aircraft owners.

Case Study #1: Runway Overrun Accident

The pilot landed a small business jet too fast and long down the runway. The airplane rolled off the end of the runway and impacted hilly terrain, while the pilot attempted to get airborne and turn to the right to avoid ground impact. Four people were onboard. Both front seat (cockpit) occupants were found wearing their shoulder harnesses but sustained serious facial/head impact injuries. Both rear-cabin occupants were uninjured. The pilot was rendered unconscious and sustained a fractured cheekbone. The right front seat passenger required facial reconstruction. Both front seat inertia reels were tested, which showed they should have properly locked during the crash event.

The government investigation agency's probable cause was "the pilot's improper decision to land with excessive speed, and his delayed decision to perform an aborted landing, both of which resulted in a runway overrun. A factor was the tail wind." Ample investigation resources and in-depth analyses were spent to accurately verify that there were *no* aircraft system anomalies that caused or contributed to the runway overrun. However, why did both front seat occupants sustain serious facial injuries while wearing shoulder harnesses?

After the government investigation agency completed their investigation and released the wreckage, the insurance company who covered both the airplane and pilot wanted to know why the serious facial injuries occurred. Examination/comparison of the subject airplane wreckage and an exemplar airplane revealed that significant aft and upward crushing of the fuselage nose section absorbed most of the impact energy, which helped prevent the occupants from sustaining fatal g-loads. None of the cockpit seats or restraint installations on the accident aircraft were changed from their originally certified configuration. Both front seat assemblies had remained attached to the cockpit floor structure, and relative displacement of the floor and seat assemblies were similar. Both control column and yoke (wheel) assemblies also remained intact and were displaced in concert with both front seats (i.e., the control column and yoke assembly did not displace closer to either of the front seat occupant's face during the crash). The overall aft crushing of the fuselage nose section was consistent with the occupants having flailed primarily forward and slightly to the right during the ground impact. Biomechanics analysis found that the serious facial injuries sustained by both front seat occupants were caused from them striking their respective control yoke.

During the exemplar airplane inspection, the accident pilot was positioned in the left front seat with its restraint system (five-point, with lap and crotch belts anchored to lower seat structure and shoulder harness inertia reeled in seat back) properly adjusted against his body. The pilot positioned his seat the same as when he was flying (adjusted

FIGURE 2.9 Accident pilot and control column/yoke positions when the pilot's face initially impacted the control yoke.

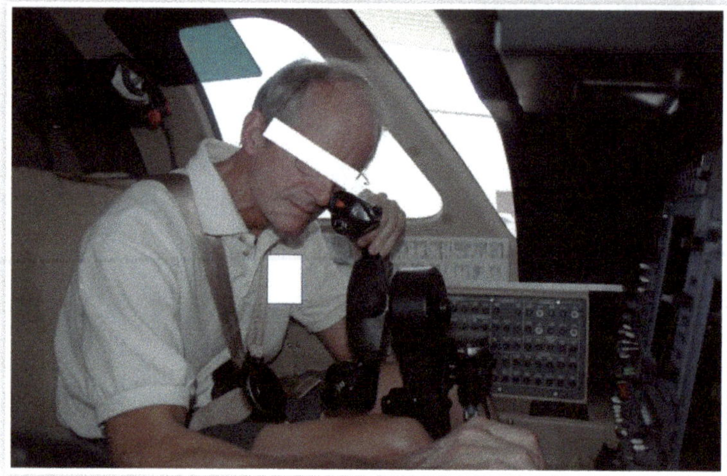

© Don Knutson

to its most aft locked position on the seat tracks, with the seat back at its most upright setting). The pilot properly adjusted his seat height and eye level above the glare shield. The shoulder harness belts were jerked forward and locked at the least possible inertia reel payout length (amount of webbing that unreeled from the seat back). Shoulder harness belt tension/length was maintained. The pilot's left hand held the control column full aft with the control wheel rotated almost 90° from a neutral setting, and the pilot's right hand was on the engine controls in the full-power position (normal positions when the pilot was trying to liftoff and turn/bank the airplane to the right to avoid impacting terrain). The pilot was then able to droop his upper body and shoulders downward (simulating vertical g-loads during a crash landing) and lean forward enough to make facial contact with the control yoke, without exerting any excessive pull force on his seat back and shoulder harness assembly (Figure 2.9).

In addition, static pull tests on the ends of both front seat shoulder harness belts were conducted. Both shoulder harness belts were jerked forward and held to the least possible inertia reel payout length. The belts were then pulled forward with 100 lb of force, and the elastic displacement of the upper portion of the seat back measured 1–5/16 in. (i.e., which would correspond to the forward motion of the pilot's upper torso); therefore, the pilot's cheek would translate forward at least another one inch into the control yoke with a shoulder harness tension force of only 100 lb. Per Federal Aviation Regulation (FAR) 23.561, the pilot should be given "every reasonable chance of escaping serious injury" during emergency landing conditions, with static inertia loads of 9.0g forward. Also, FAR 23.562 requires the seat assembly and installation to withstand peak dynamic loads of about 26g forward (with 10° yaw) and 19g downward (with 30° pitch up). This means that the pilot's upper torso would easily exceed 100 lb of forward pull force on the shoulder harness and more than 1–5/16 in. of seat back deflection during an FAA-defined dynamic crash condition.

While researching cockpit seat certification, it is interesting to note that crash sled tests were conducted with the seat and restraint assembly, an instrumented crash dummy, and a mock instrument panel and glare shield. *A control column and yoke assembly was not included because head impact with the instrument panel or glare shield, not the yoke, was assumed.* The normal flight control response for a pilot preparing to contact terrain would be to pitch/pull the nose up (control column/yoke aft) to minimize a direct head-on collision; therefore, facial impact with the control yoke should have been a practical consideration during the crash sled (i.e., aircraft certification) tests. In other words, *the shoulder harness installations were not adequate to protect the front seat occupants per the FARs.*

Case Study #2: Stall/Spin Accident

An instructor and student took off in a single-engine, two-seat trainer airplane, in good weather with full fuel. They were practicing a stall/spin maneuver from approximately 3,000 ft above ground level (AGL). Available radar data and witness statements indicated that the airplane maintained its descent until ground impact. Witnesses saw the airplane "spiraling" in a nose-down attitude. Both occupants sustained fatal injuries related to the terrain impact. The fuselage nose section and cockpit/cabin structure exhibited substantial aft and upward deformation, but remained mostly together and was resting upright. Pertinent wreckage inspection findings by the government investigation agency included the following:

- No evidence of fire or smoke in the cockpit.
- Cockpit instruments and controls were destroyed during the ground impact.
- Propeller blades did not exhibit twisting or chordwise scratches, i.e., no evidence of engine torque during ground impact.
- Wing flaps were up/retracted.
- The tail section remained intact and did not exhibit any appreciable deformation or damage to its rudder installation.
- Rudder exhibited full left deflection, with its rudder horn stop plate over-traveled and snagged under the stop bolt head (Figure 2.10).
- Rudder bumper tangs were oriented backwards (pointed aft) instead of forward as shown on Figure 2.11.
- Flight control cable continuity was established from the cockpit to the elevator, rudder, and both ailerons.
- No airframe structural anomalies were found (e.g., preexisting fractures or fatigue failures).

Maintenance records indicated that the airframe had at least 10,700 h total time, and no major repairs or alterations were performed on the rudder control system. The aircraft manufacturer issued a Service Bulletin about 3 and a half years prior to this accident, which provided an improved rudder stop installation designed to assist in preventing the possibility of the rudder overriding the stop bolt during a full left or right deflection (simply by increasing contact area of the stop bolt heads and their stop plates/bumpers). This Service Bulletin was not

FIGURE 2.10 Top view shows right stop plate snagged on its stop bolt head and the bumper tang installed backward. Bottom view shows the rudder horn with a correctly installed bumper tang.

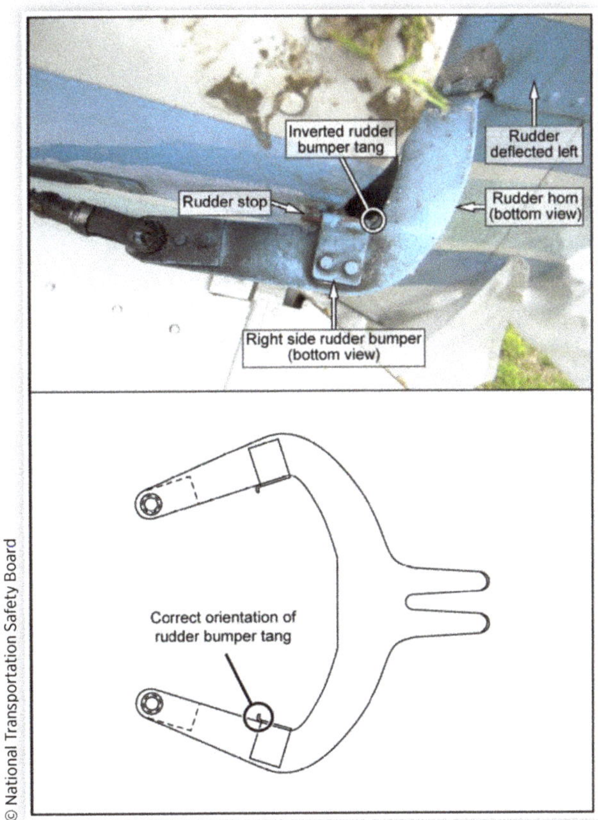

FIGURE 2.11 Exemplar rudder stop installation – view of left stop plate properly against its stop bolt during full right rudder deflection.

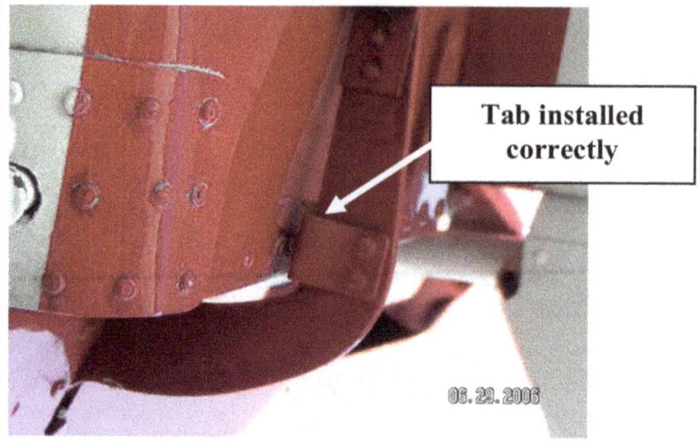

FIGURE 2.12 Rudder over-travel found on the accident in Canada.

complied with (nor was it required for continued airworthiness), and records also indicated that the rudder stop plates were never replaced or repaired.

Research of stall/spin occurrences involving a rudder being stuck in an over-traveled condition on the same airplane model revealed an accident that *appeared* to be closely related in Canada about 7 years prior to the subject accident (Figure 2.12). However, the overall circumstances surrounding the rudder over-travel was reportedly due to neglectful maintenance, which did not correlate with the subject accident airplane. In addition, the subject airplane model had been in existence for around 28 years, with over 7,500 produced, and the rudder control system did not have a history of design flaws or recurring problems.

The government investigation agency's probable cause for the subject airplane accident was an "improperly installed rudder bumper, which resulted in a rudder jam during spin training and subsequent uncontrolled descent into terrain. A factor was the operator did not comply with the service bulletin." Furthermore, a Safety Recommendation was submitted that led to an AD mandating compliance with the aforementioned Service Bulletin.

FIGURE 2.13 Cockpit engine controls showing the throttle in a full-power (forward) position and the mixture in a fuel cutoff (aft) position.

© Don Knutson

Beyond the government agency's investigation, further findings revealed evidence that did not support the Safety Recommendation and subsequent AD:

- Engine throttle and mixture controls in the cockpit wreckage exhibited positions *opposite* of what would be normally expected during a stall/spin recovery – throttle was at a full-power (forward) setting, and mixture was at a fuel cutoff (aft) setting (Figure 2.13).
- No abnormal wear or slop was found on the rudder installation such as loose hinge bearings or support brackets.
- Contact areas between both rudder stop plates and their respective bolt heads exhibited pronounced wear patterns consistent with a properly rigged rudder system (Figure 2.14), unlike reported findings from the Canada accident.
- Abrupt rudder pedal push tests (just short of damaging the pedals and rudder stop plates) on several exemplar/serviceable airplane models would not create a rudder over-travel condition nor any contact against the rudder stop plate tab (pointed forward).
- Rudder cable pull tests (>350 lb of tension applied on either the right or left control cable) on an exemplar tail section assembly mock-up/test fixture as well as the subject damaged tail section assembly would not create a rudder over-travel condition nor any contact with the rudder stop plate tab (even when the stop bolt heads were lubricated).
- In order to recreate an over-traveled condition where the rudder stop plate snagged on its bolt head (with no resulting appreciable deformation or damage to the rudder installation), the rudder assembly on both the exemplar tail section mock-up and the subject tail section had to be *forced by hand beyond its deflection/rigged limits* (while remaining within its elastic range), which could not be achieved by pulling on the rudder cables via pilot input (i.e., the only external force to aid the rudder over-travel would be from ground impact/ deceleration loads).

CHAPTER 2

FIGURE 2.14 Both rudder stop plates showing distinct and normal wear patterns, and no evidence of misalignment with their stop bolt heads.

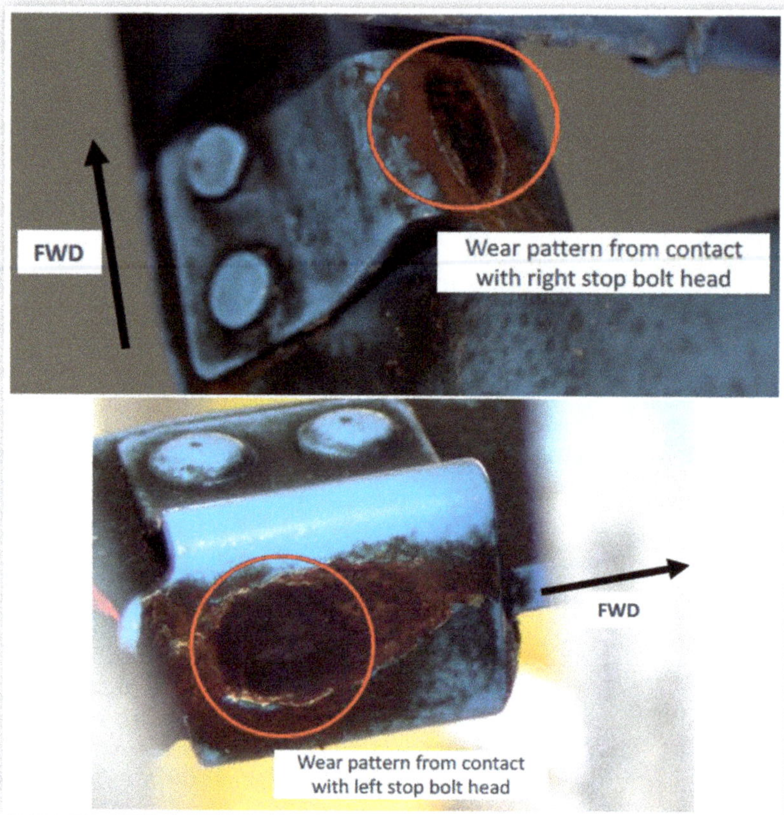

In essence, a "cut-and-paste" analysis from the Canada accident was improperly used to substantiate the subject accident findings. The above-noted tests clearly proved that *the rudder over-travel occurred during terrain impact, and something else was involved with the pilots not regaining control of the airplane* (e.g., performing the stall/spin with inadequate altitude over terrain and improper flight control inputs that created distractions that delayed the spin recovery). Also, Normal Procedures in the Pilot's Operating Handbook recommend that stall/spin entries be accomplished at a high enough altitude that recoveries are completed at no less than 4,000 ft AGL, and radar data indicated that the accident airplane entered a stall/spin at ~3,000 ft AGL.

Summary of Lessons Learned

So, practically speaking, what can we do to reduce inaccurate and incomplete investigation findings?

1. Base conclusions on physical evidence, scientific principles, and testing – not just experience, findings from related accidents, and witness statements.
2. Use the nine-box matrix (or similar technique) to prepare a more comprehensive investigative checklist *prior to* arriving on-scene or at an inspection.

3. Always look past the perceived obvious, and consider the "absurd" – things are not always what they seem.
4. Completely document and evaluate all available and detailed facts before forming any conclusions.
5. No "cut-and-paste" analyses allowed – confirm the validity of other investigation findings to ensure accurate comparisons.
6. Evaluate the injury-producing mechanisms, not just the mechanical failures – remember that the occupants on board the aircraft are an extension of the wreckage.
7. Maintain a comprehensive database, and thoroughly document both major and minor events when possible – empirical knowledge helps us evaluate trends that can lead to accidents.

Answer to cryptogram: **IF BLACK BOXES SURVIVE AIR CRASHES, THEN WHY ISN'T THE WHOLE AIRPLANE MADE OUT OF THAT STUFF?** (a quote from the famous late comedian, George Carlin)

CHAPTER 3

On-Scene Investigation

This chapter discusses a typical "field" investigation (versus a "major" investigation requiring a "Go-Team"); however, this information is still useful for major or military accidents/mishaps. Comprehensive documentation of the on-scene phase of the investigation is critical primarily because it cannot be physically preserved like the aircraft wreckage itself. This is why we should "bring the crash scene back to our office" via detailed notes, photos, videos, sketches, maps, and measurements.

Even though consultants/experts are not normally performing the initial crash site inspections, they can still benefit from this chapter because several sections are applicable to their roles as investigators, and it is important to understand what was (and was not) accomplished during the initial stages of the investigation.

Government investigation agencies are in charge of the crash investigation from the on-scene phase (when the crash occurred) until they eventually release the wreckage (e.g., to the aircraft owner or insurance company who has the hull policy). The government agency grants party status to organizations and manufacturers whom they deem necessary to provide the required technical assistance.

Accident or incident investigations that are handled without the investigator-in-charge (IIC) being present at the scene (e.g., NTSB "limited" investigations*) are typically an ineffective practice because the person designated to manage the on-scene inspection (who works for the FAA on investigations conducted in the United States) is typically not as experienced with conducting crash investigations as the IIC (i.e., AAI is not their

* An NTSB limited investigation usually occurs when the accident does not involve fatalities. NTSB investigators do not travel to the scene; instead, the on-scene phase of the investigation is delegated to FAA. The FAA forwards this information to the NTSB IIC.

primary job). This increases the chances that evidence at the crash scene will be overlooked and lost once the wreckage is recovered.

Field investigators should find as much time as practical before arriving at the accident site to plan a "to-do list." Due to time limitations before ATC data can be discarded (e.g., 15 days after the accident), the IIC will typically make sure all available subject voice recordings/transcripts and flight position information (e.g., from radar and automatic dependent surveillance-broadcast [ADS-B]*) are secured by the applicable governing agency (e.g., FAA).

Before the Investigation Team Arrived

In many cases, the accident scene and aircraft will get altered because of first responders and emergency personnel. This includes cockpit or cabin structure being breached to remove the injured, emergency vehicles driven over terrain impact marks, fire retardants sprayed on burning wreckage, fragmented wreckage being moved from its original position, and even people stealing pieces of wreckage for souvenirs. Ideally, emergency personnel should not try to move any fatally injured aircraft occupants because of potentially altering important evidence, but that is easier said than done when they do not always know whether someone can be saved. With all this said, here are some relevant questions to answer:

- Who initially arrived on-scene (e.g., witnesses, police, or firemen)?
- If a fire was involved, how was it attacked? From which direction? Type of retardant used?
- Who specifically secured or watched over the crash site? Typically, a law enforcement agency is dispatched to guard the scene, but this will not always be the case when the aircraft crashes in remote terrain.
- Was anybody seen handling the wreckage before the scene was secured? If so, who, and what did they do? I investigated a crash in the Bahamas where the propeller was stolen from the crash site and was a key piece of evidence regarding the engine power at the time of impact. The propeller was eventually retrieved by locally advertising a $700 cash reward for its return without reprisal.
- Where was each occupant found inside or outside the aircraft? If they were seated, what seat locations and orientations (facing toward the front, rear, or side)?
- How were the injured occupants removed from the wreckage (e.g., seat restraints unbuckled, seats repositioned, doors opened, cabin windows broken out, or airframe cut open)?
- Did any of the occupants possess cell phones or cameras that can be checked for photos or videos of the airplane and its occupants before or during the accident flight?
- What was the emergency personnel (e.g., fire truck and ambulance) response times from when the accident occurred?

---

* ADS-B requirements are covered in 14 CFR 91.225 and 91.227, and the following are some advantages noted by the FAA:

- Enhances safety and efficiency and provides greater coverage by moving from ground radar and navigational aids to precise tracking using satellite signals.
- Pilots can see what controllers see – displays showing other aircraft in the sky and hazardous weather and terrain.
- Reduces the risk of runway incursions with cockpit and controller displays that show the location of aircraft and equipped ground vehicles on airport surfaces – even at night or during heavy rainfall.

- If wreckage was removed from salt or brackish water, was it rinsed with fresh water to reduce the corrosive effects on unprotected metals?

The above-noted information should be obtained as soon as possible, including photos, video, and sketches. As time passes, the accuracy of the emergency responders' memory will be compromised just like any witness, especially when events are not comprehensively documented.

Basic Investigation Gear

As already indicated in Chapter 1, when conducting the on-scene investigation, we typically will not have the luxury of inspecting the aircraft in a climate-controlled hangar nor be provided door-to-door travel arrangements to the accident site. We can easily end up in a remote area - on the side of a mountain or in the middle of a swamp, desert, or large body of water. Therefore, we need to be mobile and flexible with the equipment we carry. We handle these diverse situations by putting together the following "go-bag:"

Technology affords us the ability to carry all the applicable airframe and powerplant technical information on a lightweight laptop computer or tablet, such as:

- POH and airplane flight manual
- Maintenance manuals (including aircraft dimensions, inspection forms, overhaul manuals, and electrical diagrams)
- Illustrated parts catalog
- ADs* and SBs†
- "Homemade" inspection data (mechanical actuator extensions corresponding to specific flight control deflections, distinct part configurations on various aircraft models, etc.)

Other useful electronic equipment includes:

- Digital camera with interchangeable lenses for both close-in and wide views, flash unit, and video capability
- Pocket/digital voice recorder
- Handheld GPS (global positioning system) (some cameras have this as well)

Other needed tools include:

- Notebook with pencil/pen
- Flashlight
- Inspection mirror
- Basic small hand tools - screwdriver (magnetic tip with flat and phillips heads), knife, pliers, etc.
- 100-foot tape measure (lightweight) and 6-in. ruler

* Airworthiness Directives (ADs) are legally enforceable rules (requiring mandatory compliance) issued by the FAA or other government regulatory agency to correct an unsafe condition in a product (an aircraft, aircraft engine, propeller, or appliance).

† Service Bulletins (SBs) are notices to aircraft operators from a manufacturer notifying them of a product safety improvement. In many cases, the SBs are the precursor to ADs.

FIGURE 3.1 Simple inclinometer using a weighted string and a straw to look through.

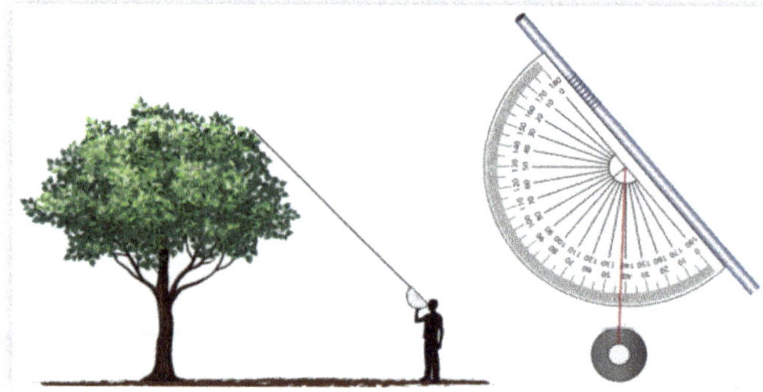

©Atoosa (https://math.stackexchange.com/users/297503/atoosa), Determine the height of the tree, URL (version: 2016-01-06): https://math.stackexchange.com/q/1602683

- Two dice for identifying right and left (green and red, respectively) piston engine cylinder components in photos
- Magnetic compass
- Compact binoculars
- Angle finder or protractor with weighted string attached (Figure 3.1)
- Magnifying glass and/or compact microscope (10X-100X)
- Plastic bags and identification tags (preserving small pieces of evidence)
- Extra batteries and chargers for aforementioned electronic equipment
- Aircraft model (not always practical)
- Applicable maps - road, topographical, airport, or aeronautical (can also be in electronic format)

Protective wear should be used such as:

- Leather gloves
- Disposable coveralls
- Biohazard gear
- Waterproof/insulated boots
- Full-brim hat (whether you're bald or not)
- Safety goggles/glasses
- Rain suit or poncho

Other safety items include:

- Alcohol wipes
- Sunscreen lotion and lip balm
- Insect repellent
- Snakebite kit
- Toilet paper
- Compact first-aid kit (Band-Aids, antiseptic, etc.)
- Water and quick energy foods

Do not think the above-noted is just another list of theoretical “stuff” – these items have been battle-tested through years of experience. Definitely pay attention to the protective wear and other safety items. Also keep in mind that all these items should be contained within a backpack that can take a beating while keeping everything clean and dry. Through experience, we learn to be more efficient with the equipment we carry, that is, less weight and bulk.

Safety Considerations

Whatever the weather conditions, investigators primarily need to protect their skin from cuts, sunburn, and frostbite. Regardless of hot or cold temperatures, always stay dry and hydrated. You cannot get immediate treatment when you are in the middle of nowhere! Also, remember that no matter how warm we dress, having cold fingers or toes makes for a miserable time. Other hazards to be on-guard for include the following:

- Sharp, jagged material on wreckage
- Composite fibers
- Shifting of wreckage (e.g., on steep slopes or deep snow)
- Pressurized systems – pneumatic, hydraulic, and oxygen
- Blowout or explosion of damaged landing gear tires
- Loaded guns or military ordinance
- Hazardous chemicals or radioactive materials
- Blood-borne pathogens (e.g., hepatitis and HIV)
- Critters (e.g., snakes and spiders)
- Undischarged ballistic recovery system cartridges or ejection seat rocket motors

Good old-fashioned planning and common sense can prevent injuries. ALWAYS THINK BEFORE YOU ACT – if you are not sure how something works or what it contains, then seek help to properly deal with it.

Emotional Considerations

Besides the aforementioned physical challenges, we must always be mentally “on-guard” when investigating catastrophes and destruction. Seeing dead bodies or parts of bodies, working long hours with a rigorous and pressure-packed schedule, being away from your loved ones, and dealing with bereaved family and friends of the deceased put very real emotional stress on us. We typically forget that our subconscious processes stress without us even realizing it. With this said, here are a few simple and healthy ways to deal with our emotions (which is sometimes easier said than done):

- Do not become “attached” to the tragedy nor keep depressed feelings to yourself – always make time to talk and maintain a sense of humor with co-workers (this does not mean you are being uncaring).
- Maintain proper rest, exercise, and diet.
- When traveling, maintain daily contact with your loved ones.
- Avoid or minimize unnecessary contact with bereaved family and friends.

Accident Scene Documentation

As addressed earlier in this chapter, a field investigator's goal should be to bring the accident site back to his/her office. Try to document everything as thoroughly as possible because a lot of evidence is lost once the wreckage is cut and disassembled during recovery. Also, trampled and cleared terrain during recovery operations, vegetation growth, and changes in season will substantially change the accident scene as you originally found it.

Prior to getting into a photo-shooting frenzy, the investigation team should conduct a visual survey of the entire scene. This will serve as a documentation planning session as well as provide an opportunity to evaluate site safety issues. Hiring a survey crew and/or an aerial photographer (e.g., using a drone or helicopter) should be considered when the wreckage scatter is substantial or the terrain is difficult to navigate, such as mountainous or densely forested areas. Hiring underwater salvage specialists should be a must when dealing with submerged wreckage.

Comprehensively document the terrain and obstacles impacted. This includes detailed descriptions of ground scars (e.g., depth, width, length, direction, and where the dirt rim around the impact crater is piled the highest from ground plowing), propeller slash mark spacing (which is further discussed in the next chapter), and diameters and orientation of broken and cut tree limbs.

Flight control trim tabs (e.g., on an aileron, elevator, stabilator, and rudder), wing flaps, and landing gear actuator positions and extensions should be documented *before* any repositioning of airframe wreckage or potential pulling on control cables. Actuator positions can be referenced from maintenance manuals, engineering drawings, or measurements taken from an exemplar aircraft installations. Rod/jackscrew extensions are commonly measured from where the rod extends out of the actuator housing or actuator support structure to the centerline of the rod end attachment (as shown in Figure 4.28 in Chapter 4). All actuator positions and extensions should be compared with their respective cockpit control handle/lever setting and position indicator (when applicable). Undisturbed actuator positions and extensions can provide evidence of pre-impact flight situations, such as an attempted emergency landing configuration, a twin-engine airplane losing power on one engine, and the pilot trimming the rudder to compensate for an asymmetric yaw condition or an inadvertent excessive pitch trim setting.* Mechanical flight control trim tab system installations using control cables and/or chains and sprockets are typically sensitive to being altered when the airframe sustains substantial impact damage or fragmentation during a crash.

Typical wreckage or crash scene diagrams will depict locations of the scattered aircraft components in rectangular (X and Y) coordinates as shown in Figure 3.2 or polar (radial and distance) coordinates as shown in Figure 3.3. Keep in mind that more than one diagram may be needed in that documenting wreckage scatter can be similar to photographic techniques – an overall view along with more detailed shots. Direction of flight and ground

* Per Chapter 6 of the *Pilot's Handbook of Aeronautical Knowledge* (FAA-H-8083-25B), trim tab systems are used to relieve the pilot of the need to maintain constant pressure on the flight controls. A mechanical trim tab system usually consists of a cockpit control wheel, two control cables (sometimes include chains and sprockets) with turnbuckles (for tension adjustments) and pulleys (for routing through airframe), an actuator with an adjustable rod end/jackscrew, and a small hinged trailing edge surface attached to a primary flight control surface (e.g., elevator, rudder, and aileron). Rotation of the cockpit control wheel results in deflection of the trim tab, which helps minimize a pilot's workload by aerodynamically assisting the movement and position of the primary flight control surface.

FIGURE 3.2 Sample wreckage distribution diagram using rectangular coordinates (objects are not to scale).

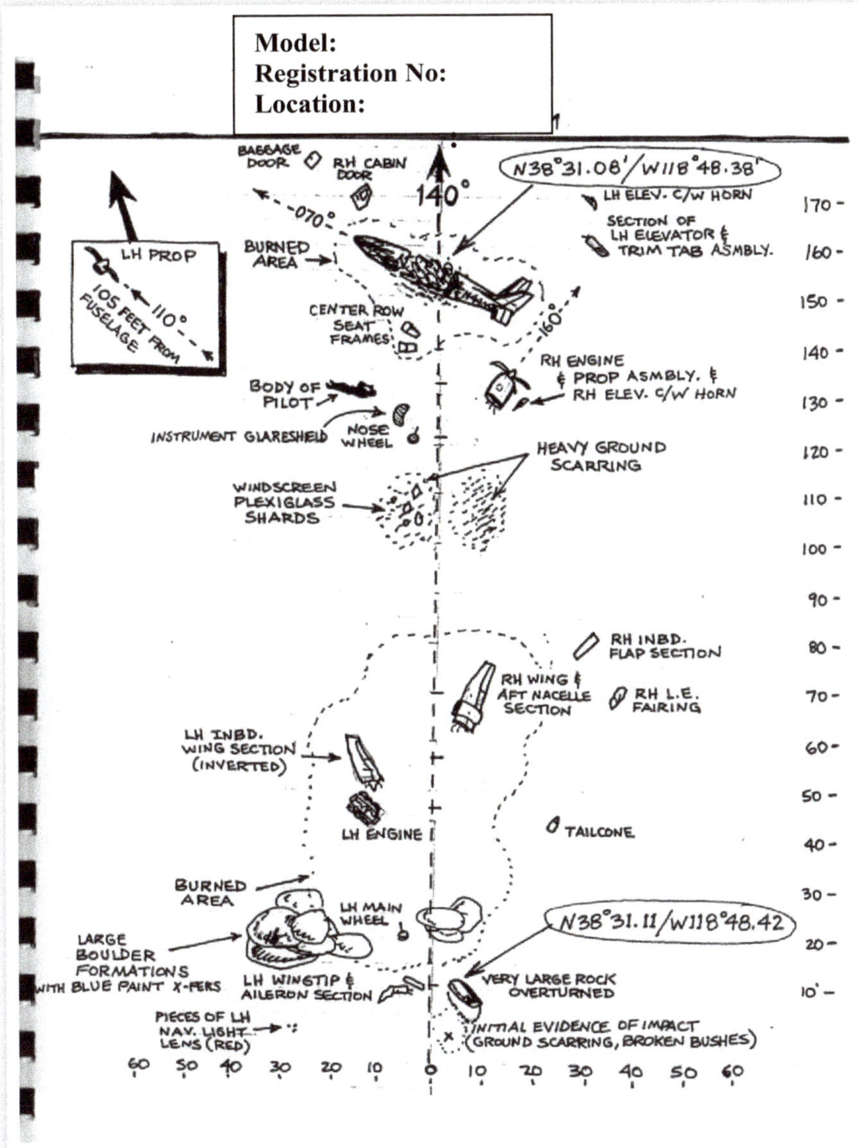

travel and locations/measurements of impact marks and obstacles are also commonly noted. The following are some other helpful pieces of information for a wreckage diagram:

- Latitude and longitude of points of interest
- Ground conditions (e.g., plowed, muddy, flooded, rocky, snow covered, burned, and fuel spills)
- Wreckage dimensions*
- Airframe crush distances and angles*

* Refer to three-view drawings of the aircraft, with fuselage stations, butt lines, and waterlines – print a copy from the maintenance manual to draw and take notes on.

FIGURE 3.3 Sample wreckage distribution diagram using radial and distance coordinates.

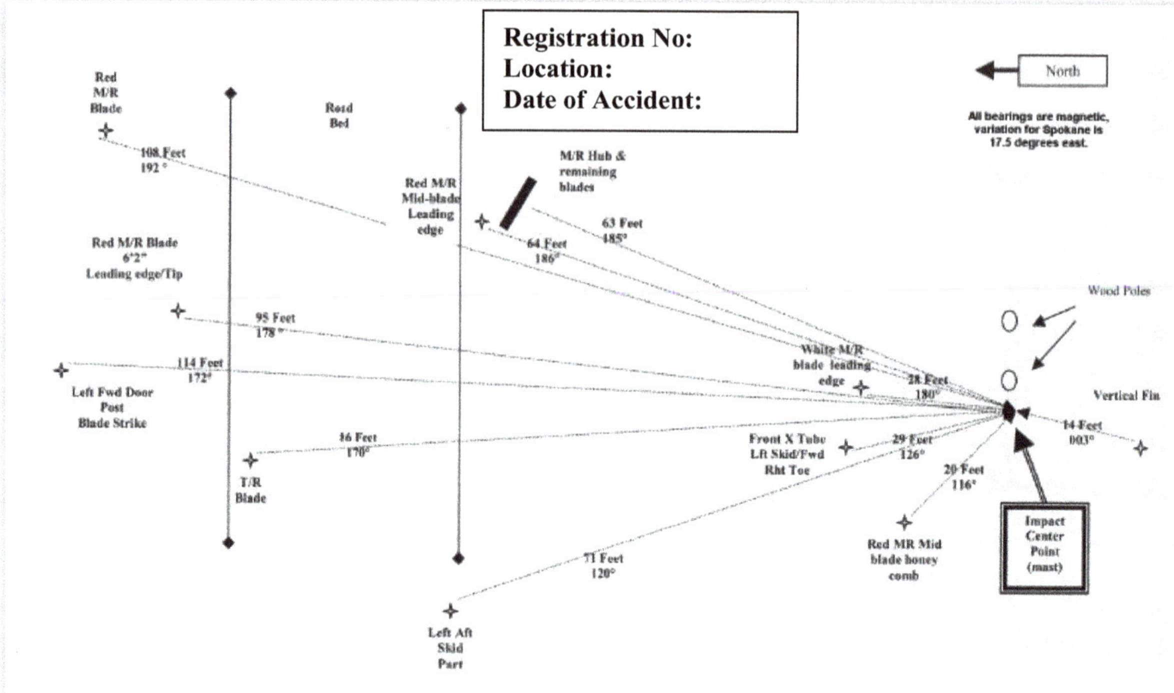

- Locations of where bodies were found
- Witness locations
- Component part numbers and serial numbers
- Referenced photos
- Referenced maps or overlays
- Topography or terrain elevations

Wreckage Layout and Recovery

Wreckage layouts *at the crash site* help determine further detailed examination requirements and accounting for components found or missing. This stage of the investigation should *not* include detailed component teardown inspections because evidence can be much better preserved and examined at a clean, organized, and properly equipped lab or designated inspection facility. Components that can be practically disassembled on-scene (with permission of the IIC and after properly documenting the assembly) include inspection panels, cowlings, fairings, and other various "quick and easy" components (e.g., engine spark plugs and valve covers to perform cylinder compression checks and verify gear/valve train continuity), fuel and oil samples, and anything deemed to be potentially damaged or lost during the recovery.

Damage to the airframe wreckage can and does occur during its recovery phase, and follow-up wreckage inspections in many cases bring up questions about whether specific damage occurred before or after the accident. Therefore, investigators should consider the following:

- Technical oversight and advice should be given to the recovery team when disassembling or cutting into the wreckage and minimizing secondary damage to the airframe during sling load operations.
- All control cables, wiring, hoses, and tubing that are cut or disassembled should be labeled (which in many cases is not done).
- Applicable fuel, oil, and debris samples should be taken prior to loading.
- Clean or wash off corrosive liquids or chemicals (e.g., salt water and fire retardants) in contact with the wreckage and fractured surfaces.
- Put instruments or avionics with electronic memory that can be sensitive to static discharge and moisture damage in sealed static shielding bags.
- Have all of the above-noted tasks photographed and/or videotaped.

Ensure pieces of interest are properly marked or tagged. Detailed descriptions of components include the following:

- Orientation or position – right, left, top, bottom, forward, aft, interior, or exterior
- Part numbers and serial numbers
- Location of part (e.g., latitude/longitude coordinates)
- Date of documentation
- Aircraft registration number and serial number

Exemplar Aircraft Inspections

Ground and flight testing of exemplar systems and components help determine how the subject system or component would function or be controlled by the pilot and the overall flight characteristics and performance capabilities prior to the crash. Some examples are evaluating aerodynamic force feedback on the flight controls (e.g., from trim tab settings), the pilot's visual reference points for a mid-air collision, cockpit/cabin volume and occupant egress issues for crashworthiness analysis, and locating possible ignition sources for fires and explosions.

Tips on Photo Documentation

During on-scene and subsequent examinations, photography is one of the main tools of our trade and sometimes requires us to be creative in order to capture as much detail as possible within each picture (i.e., "A picture is worth 1,000 words."). Therefore, it is imperative that we learn how to properly use our camera. Here are some *basic* recommendations:

- Take some macroscopic views of the object's location relative to other parts, and then progressively focus in more on the object itself - shoot at least an overall, midrange, and close-up.
- Objects are three-dimensional, so shoot all necessary sides and angles.

FIGURE 3.4 Oblique or corner view of lower forward section of a fuselage cabin and engine firewall, which is good for showing the overall damage.

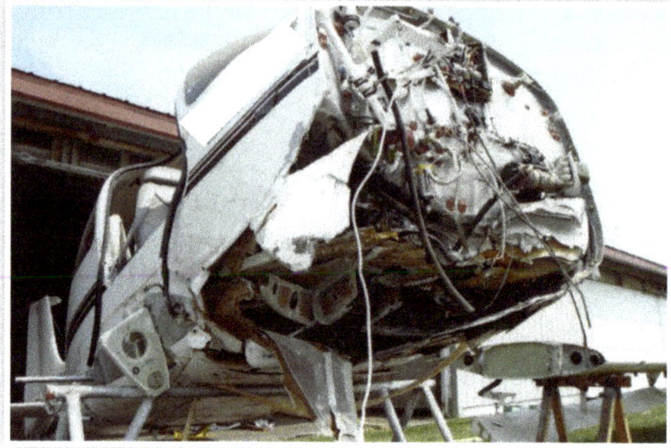

- Flash units are usually needed indoors, during low-light situations, or when filling in shadows within wreckage (such as background objects inside of crumpled structure or within a cowling).
- Include a ruler or other dimensional reference with the object being documented.
- Use a straightedge to emphasize deformation of surfaces.
- Take oblique or corner photos to document the *overall damage or condition*, and direct front, side, top, etc. photos to depict the *amount and direction of deformation* (Figure 3.4, Figure 3.5, and Figure 3.6).
- Include a color chart/reference (for digital photo editing) when the accuracy of colors are important, such as emphasizing the discoloration of surfaces.
- Use views that show direction of damage by relating one assembly to another (Figure 3.7).
- Use a mirror to show features that are hard to access and behind other objects such as components inside an inspection panel or heavily populated engine compartment.

FIGURE 3.5 Direct front view of forward section of a fuselage cabin and engine firewall, which is good for showing amount and direction of damage.

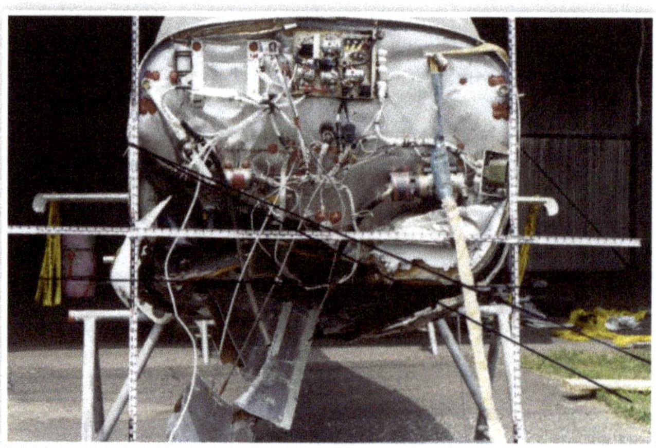

FIGURE 3.6 Direct side view of lower forward section of a fuselage cabin and engine firewall, which is good for showing amount and direction of damage.

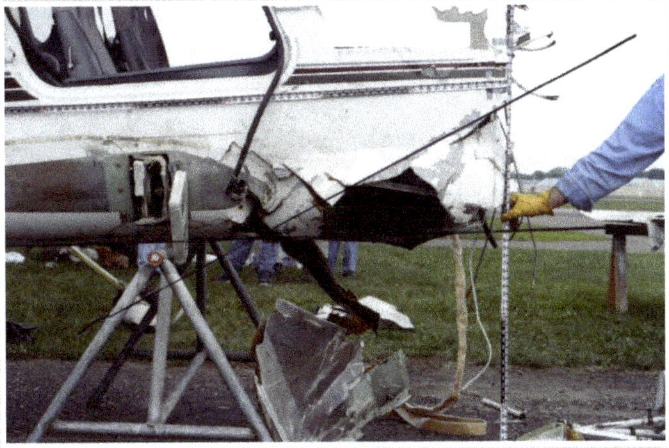

© Don Knutson

FIGURE 3.7 Chordwise scratches on propeller blade in relation to propeller spinner and forward engine cowling, which clearly shows propeller rotation during terrain impact.

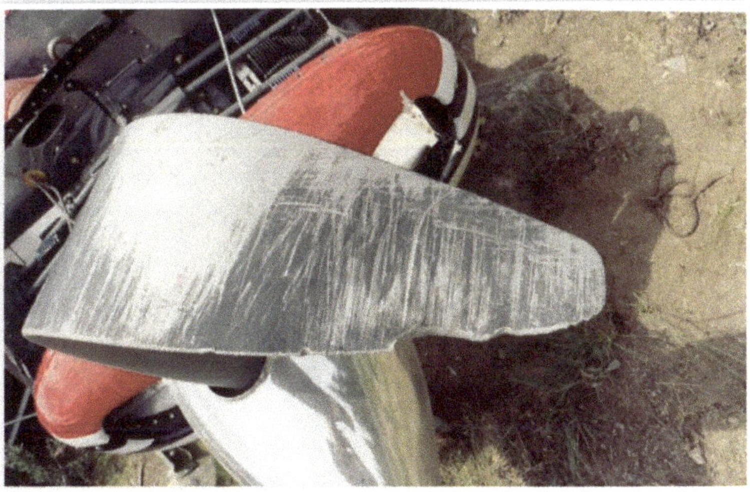

© Don Knutson

- Turn/look around – some photos may be impossible to relate the location of the wreckage within the given terrain features without additional photos of the surroundings.
- Consider exposing multiple overlapping images, which can be taped together to form a panorama (remember not to zoom between shooting these images) – digital cameras typically have a panoramic setting.
- A three-dimensional image can be produced by taking two photos focused at the same distance and laterally apart. The lateral distance depends on how far the camera is from the object (e.g., 12 in. laterally apart when 100 ft between the camera and object, 3 in. when 6-9 ft, and 0.5 in. for close-ups). A stereoscope is then used to allow each eye to focus on one of the photos simultaneously.
- **Do not worry about taking too many photos!**

Occupant Injuries

In addition to gathering all the emergency responders' information, the IIC (on behalf of the investigation team) should ask the medical specialists performing the autopsies and toxicological exams, and providing hospital care, to look for specific evidence that is suspected to be related to the crash.*, † Examples include the following:

- Levels of carboxyhemoglobin in blood from carbon monoxide leaking into the cabin/cockpit because of a shrouded muffler or combustion burner can used in a cabin heater system being cracked.
- Trachea or other airways exhibiting soot showing the occupant was alive at the time of exposure to the in-flight or post-impact fire.
- Bruises on upper body that are consistent with occupants wearing a shoulder harness during the crash.
- Facial bone fractures consistent with the occupants flailing against the control yoke or instrument panel.
- Hand and foot injuries that are consistent with the pilot attempting to control the aircraft (versus being incapacitated) or who was controlling the aircraft when two pilots were seated in the cockpit (this evaluation should include computed tomography (CT) scans‡ to evaluate impact trauma such as dislocations, fractures, and tears).
- Spinal compression injuries to estimate the aircraft vertical impact load and velocity.

Remember that bodily injuries are related to and an extension of the aircraft wreckage. Research on human tolerance to injuries is plentiful, which means that this empirical knowledge can be readily applied to help quantify and qualify impact forces. Knowing the pilot's smoking and drinking habits, eyesight, history of injuries, surgeries, etc., can also help evaluate circumstances contributing to the accident.

Witness Interviews and Statements

As already touched upon in the previous chapter, our memories are inefficient and become even more so as time passes; therefore, interview the witness as soon as possible after the accident. Here is some helpful advice:

- Ensure the witness is as comfortable as possible – empathize with their situation.
- Keep other witnesses separated during interviews as much as possible – encourage witnesses not to discuss the details of the incident with each other before their statements are taken (good luck).

---

* Handy references for aviation-related medical and human performance issues include FAA Aerospace Medicine Technical Reports (online); and Fundamentals of Aerospace Medicine, Fourth Edition, ISBN 9780781774666.

† Some good introductions to the evaluation of crash injuries are NATO Research and Technology Organisation publication number RTO-EN-HFM-113, "Pathological Aspects and Associated Biodynamics in Aircraft Accident Investigation" (2005); and Advisory Group for Aerospace Research & Development (AGARD) Lecture Series 208, "Injury Prevention in Aircraft Crashes: Investigative Techniques and Applications" (1998), ISBN 92-836-1068-7.

‡ Computed tomography (CT) scans are procedures that use a computer linked to an X-ray machine to take a series of detailed pictures of areas inside the body. These pictures are taken from different angles and used to create three-dimensional views of bones and joints and soft tissues, such as cartilage, muscles, and tendons.

- Encourage the witness to recall as much detail as possible – encourage them *not* to guess or "fill in the blanks."
- Ask open-ended questions (e.g., "What can you tell me about the airplane?"), followed by closed-ended questions *after* they are done (e.g., "What color was the airplane?").
- Once a question is asked, let the witness finish answering without interruption.
- "I don't remember" *is* an acceptable answer – do not push (e.g., "Are you sure?"), otherwise the witness may feel compelled to fill in the blanks.
- Provide the witness with an aircraft model and/or other visual aids.
- Explore the witness' areas of expertise and what they know about aviation.
- If possible, allow the witness to be at the place he/she witnessed the event.
- Summarize for the witness your understanding of their account.
- Document the witnesses' statement in their own words – do not paraphrase.

Typical challenges with witnesses include their perception based on lack of experience (e.g., difficulty describing an airplane part due to being unfamiliar with aircraft in general), optical illusions (e.g., Boeing 747 looks like it's flying 20 MPH when you're standing five miles away), quality of the interview process (e.g., leading questions or intimidation), and experience with and relationship to the people involved in the accident (e.g., "he/she was the safest pilot I ever met").

Understanding the News Media

Our society obviously benefits from staying informed, and a news reporter's job is understandably to find answers. However, the news media also can have a tendency to sensationalize events to "sell" their news, which means that reporters are not always going to be factually or technically accurate. When it comes to AAI and the news media (television, radio, newspapers, magazines, web pages, blogs, etc.), talking to reporters about an accident should be left to the government agency in charge of the investigation. A field investigator should always remain cognizant of any news reporters present during a field investigation and avoid any discussions with them either directly or in earshot. Even a known off-camera or "unrecorded" conversation should be considered taboo. I learned this firsthand as an investigator for Beechcraft while investigating a notable airplane crash on a mountain side. While hiking back down the mountain after examining the wreckage, I walked by a woman dressed in plain casual clothes and a camera who simply asked me what happened. Not fully paying attention to *whom* I was talking with, my response was brief and that the airplane somehow lost power and was forced into an emergency landing. The woman asked me who I was and I said I worked for the airplane manufacturer. That is all I said, and then continued down the mountain … and guess what? While picking up a popular newspaper at the hotel during breakfast the following morning, lo and behold, there's a photo of the airplane wreckage on the mountainside with a caption to the effect of *Beech investigator says airplane ran out of fuel* (and of course, I was the only Beech person who had any contact with the media).

CHAPTER 4

"Kickin Tin"

This chapter discusses *basic methods and techniques* regarding how to "read" wreckage and "visualize" a crash event. Actual aircraft accidents and component failures are used as examples to provide general investigative concepts that can be readily applied to other aircraft and components with similar designs, materials, and configurations. These concepts include how to evaluate [i] aircraft orientation during a crash; [ii] sequence and modes of airframe failures; [iii] interactions and relative movement of aircraft components during impact, deformation, and separation; and [iv] some circumstances that possibly existed during a crash such as flight control and engine power settings, systems operation, or who was on the controls. Further wreckage reconstruction methods and techniques will be discussed in subsequent chapters that dovetail into this chapter.

Wreckage Distribution

In many cases, wreckage diagrams will show parts scattered in a pattern that gives us a basic idea of how the aircraft initially impacted terrain, which is typically the genesis of wreckage reconstruction. The following are some general ground impact conditions along with the resulting wreckage scatter and ground scars (and assuming the crash occurred in a relatively open area):

- High-speed (e.g., at or above cruise speed or during a spiraling descent) ground impacts in a steep nose-down attitude typically show parts spread radially outward from the initial point of impact. The extent of the ground impact crater or imprint and wreckage scatter and fragmentation depend on the condition of the terrain (tilled or packed soil, pavement, muddy, etc.). The approximate horizontal direction of flight during terrain impact can sometimes be determined from the highest point of the crater's rim or how the terrain has been mostly pushed outward and upward.

- High-speed ground impacts while in a shallow descent typically result in substantial wreckage scatter distances, and create a fan-shaped wreckage distribution pattern. In many cases, the heavier parts that readily detach from the airframe during impact can end up traveling the farthest (e.g., the engine[s]). Airplanes impacting the ground in this configuration with a high bank angle usually cartwheel when clipping the outboard portion of their wing, which typically results in additional airframe fragmentation.
- Low-speed (e.g., near or in a stalled condition) ground impacts in a steep nose-down attitude usually create a shallow crater or ground imprint with very little wreckage scatter.
- During an airplane stall/spin, a helical/vertical descent path is maintained until ground impact, which typically results in limited wreckage scatter (Figure 4.1). The initial/lowest wing that contacts terrain usually sustains more deformation such as leading edge aft crushing or structural bending, which commonly indicates the direction of spin. Also, the side of the aircraft that exhibits the most aft displacement from the initial impact ground scars usually indicates the direction of the spin (Figures 4.2 and 4.3). The referenced figures show the left engine and its wing nacelle displaced notably aft of the initial left propeller impact and the right propeller displaced forward into the ground; therefore, the airplane was in a left spin. The greater the nose-down attitude, the more aft crushing or displacement will usually be found on the fuselage nose section, engine mounting, and/or wing leading edges.

FIGURE 4.1 Airplane wreckage after a "flat" spin accident.*

© Don Knutson

* NASA categorizes spin modes as steep (20-30°), moderately steep (30-45°), moderately flat (45-65°), and flat (65-90°). A good reference concerning stall/spin situations is The Light Airplane Pilot's Guide to Stall/Spin Awareness, by Rich Stowell (2007), ISBN 9781879425439.

FIGURE 4.2 Left propeller assembly displaced aft of its initial ground impact (note soil imprint from propeller blades).

© Don Knutson

Airframe Structural Damage

Specialists in the materials field who can effectively assist wreckage reconstructionists include metallurgists, chemists, tribologists, stress engineers, and nondestructive testing (NDT) experts. Typical information provided by these specialists include:

- Material properties and composition
- Instantaneous versus progressive or fatigue failures (e.g., possible preexisting material flaws)
- Load cycles before failure or designed life limit (e.g., aging aircraft issues)
- Manufacturing processes (e.g., heat treatment of metals, composite layups, moldings, castings, welds, and machining)
- Fire and explosion analysis (e.g., cause and origin and deflagration versus detonation)
- Wear limitations and lubrication requirements (e.g., engine components and actuator jackscrews)
- Load paths and stress within airframe structures (especially ones that are complex such as tubular engine mount or truss fuselage assemblies) to help evaluate the sequence or weak link of failure (e.g., using finite element analysis [FEA]* as shown in Figure 4.4)

FIGURE 4.3 Right propeller assembly displaced forward of its initial ground impact (note spinner driven into grassy terrain).

© Don Knutson

* FEA is a computerized modeling method for predicting how a structure reacts to real-world forces, vibration, heat, fluid flow, and other physical effects. This advanced engineering tool can augment/replace experimental testing.

FIGURE 4.4 FEA dynamic modeling of an engine mount and nose landing gear assembly during terrain impact, with color-coded stress values.

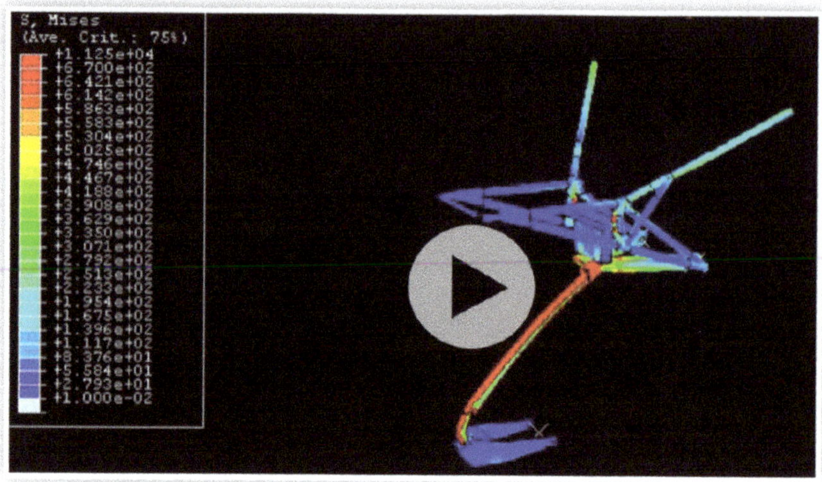

© Don Knutson

Most primary airframe structures* are built with aluminum and steel alloys and composites with fiber-reinforced polymers. Aluminum airframe structures (fuselage, wings and tail section) in a crash typically result in pronounced or crushable damage signatures, which usually makes them easier to "read" than composite airframe structures. Figure 4.5 provides a good example of an airplane's aluminum wing structure exhibiting chordwise/aft crushing consistent with at least a 70° nose-down attitude during ground impact. Composite airframe structures with fiber-reinforced polymers are generally rigid to the point where they can exhibit little or no deformation prior to ultimate failure.

Evaluation of overall airframe damage from crash and aerodynamic loads includes documenting the direction of impact forces, amount and orientation of deformation,

FIGURE 4.5 Chordwise views of both wings of an airplane exhibiting extensive crushing damage.

© Don Knutson

* Primary structure supports the aircraft's flight and ground loads, and its failure would reduce the structural integrity that can result in injuries to the crew and passengers.

separated or missing parts, and fire consumption. Angles and contours of crushed primary structures help determine the orientation of the aircraft during impact. Crush angles *and* distances on primary structures help estimate impact forces or g-loads (crashworthiness and survivability analysis is covered in Chapter 5). Aircraft station diagrams and one-, two-, three-view drawings as shown on Figures 4.6 and 4.7 help clearly map this out. Common airframe crash or overload damage includes:

- Aft and *upward* crushing of the forward airframe structure (e.g., fuselage nose section, engine mounts, and wing leading edges), which is consistent with terrain impact in an *upright* attitude. An *inverted* aircraft impact attitude typically results in the forward airframe structure exhibiting aft and *downward* crushing.

FIGURE 4.6 Diagram showing fuselage stations (FS), waterlines (WL), buttock lines (BL), and wing stations (WS).

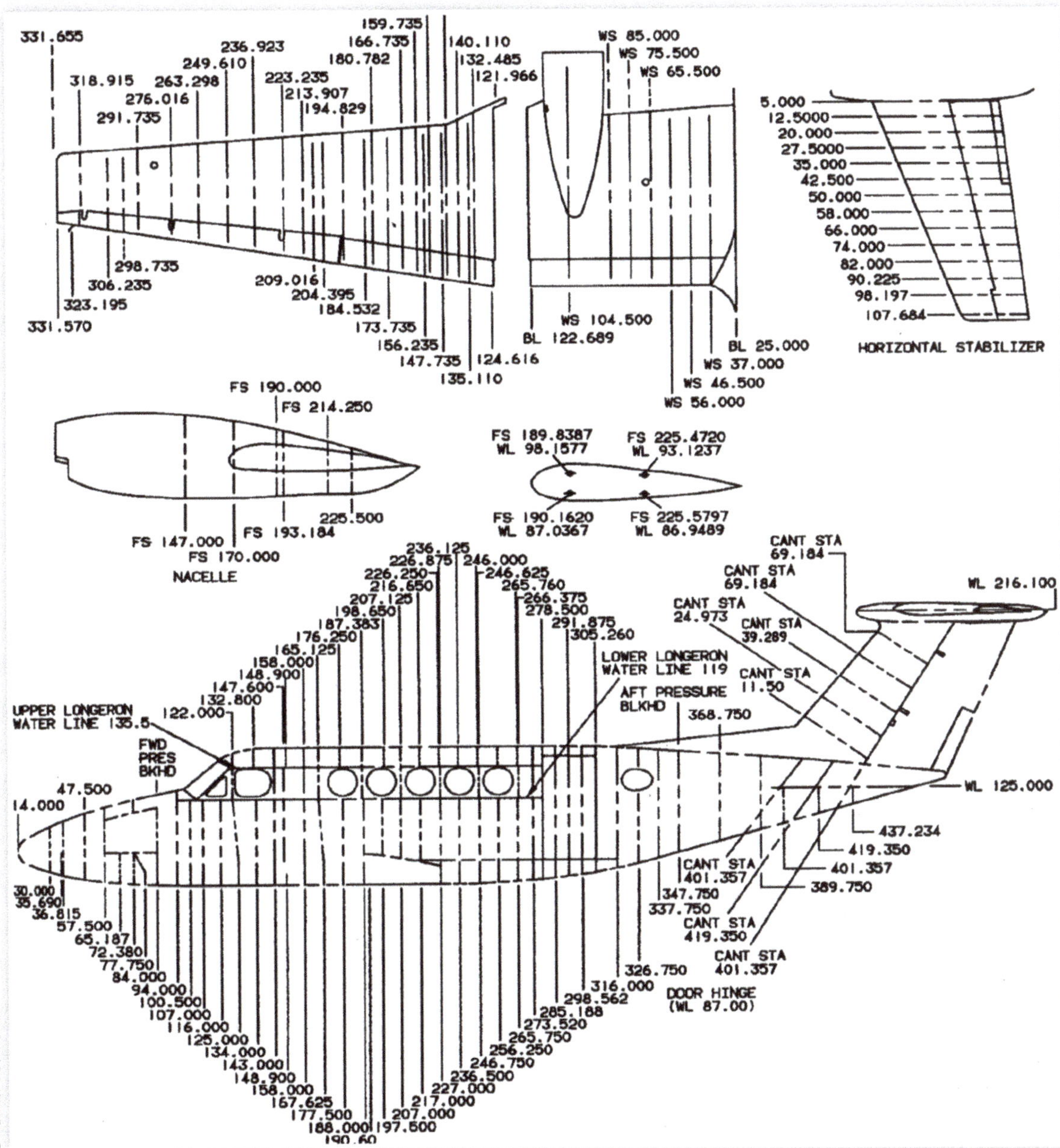

FIGURE 4.7 Three-view of an airplane involved in an in-flight breakup, using numbered references for the fragmented parts.*

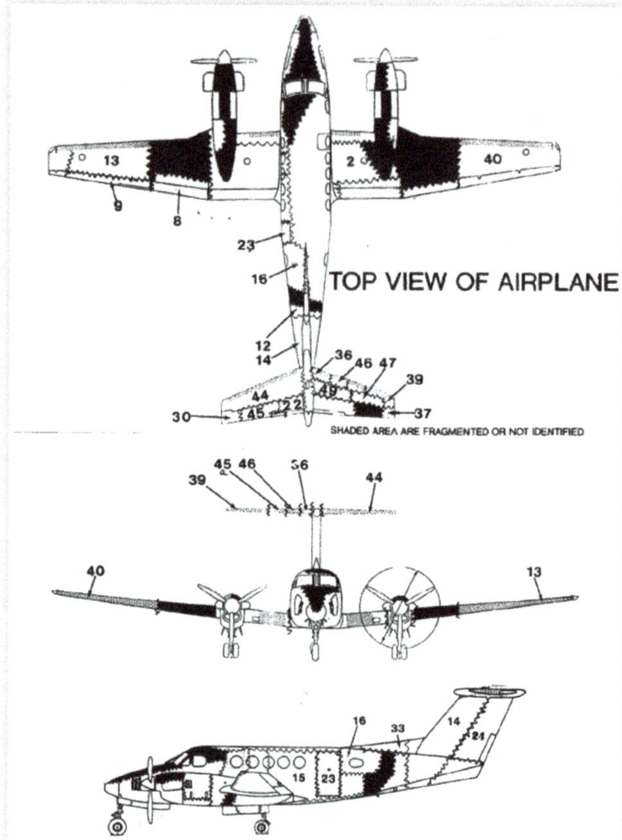

- *Diagonally* buckled/wrinkled aluminum skin panels between bulkheads and longerons on fuselages and ribs and spars on wings and stabilizers (Figure 4.8), which are consistent with a combination of structural *twisting and bending* modes. Vibrations driven by aerodynamic forces can also cause this type of damage.
- *Lateral* buckling of aluminum skin panels between bulkheads/longerons on fuselages and chordwise buckling between ribs/spars on wings and stabilizers, which are consistent with structural *bending* modes (including aerodynamically driven vibrations).
- Slap and scratch marks, paint transfers or smears, and impact indentations or punctures, which occur between parts that are close-fitting or detaching (e.g., door swung open or forced against its fuselage frame and in-flight separation of a wing that impacted the tail section).
- Debris from terrain such as soil, tree branches, or leaves that can get trapped in deformed or torn open structure.

Figure 4.9 shows an example of mapping and visualizing overall impact forces imparted on an aircraft during a crash. Blue arrows represent inward collapsing of the lower fuselage cabin structure adjacent to the wings. Black arrows represent downward collapsing of the fuselage cabin structure. This technique also shows how specific parts

* NTSB/AAR-03/01, In-Flight Electrical System Failure and Loss of Control, Jet Express Services, Raytheon (Beechcraft) Super King Air 200, N81PF, near Strasburg, Colorado, January 27, 2001.

FIGURE 4.8 Example of buckled aluminum skin panel found on wings.

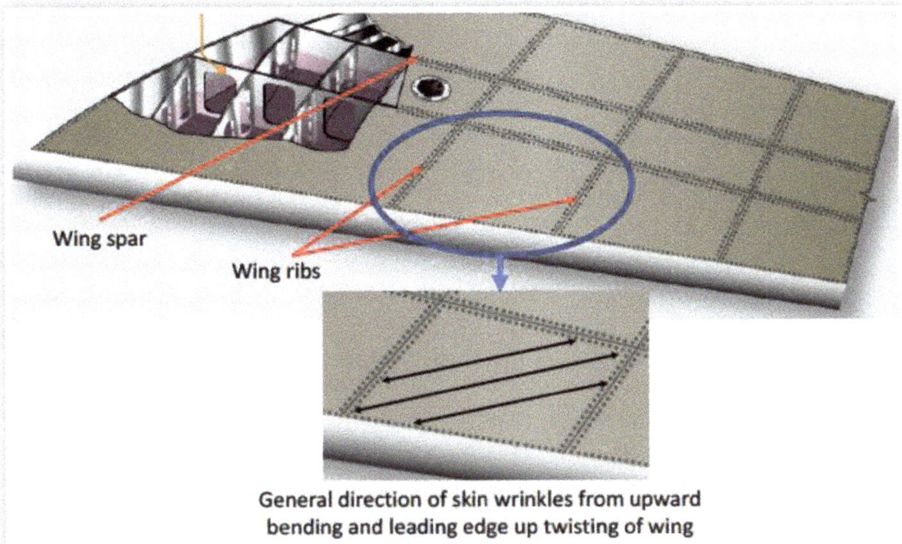

FIGURE 4.9 Beech A36 Bonanza three-view diagram showing directions of initial impact forces (red arrows) and resultant direction of deformation to the fuselage cabin (blue and black arrows).

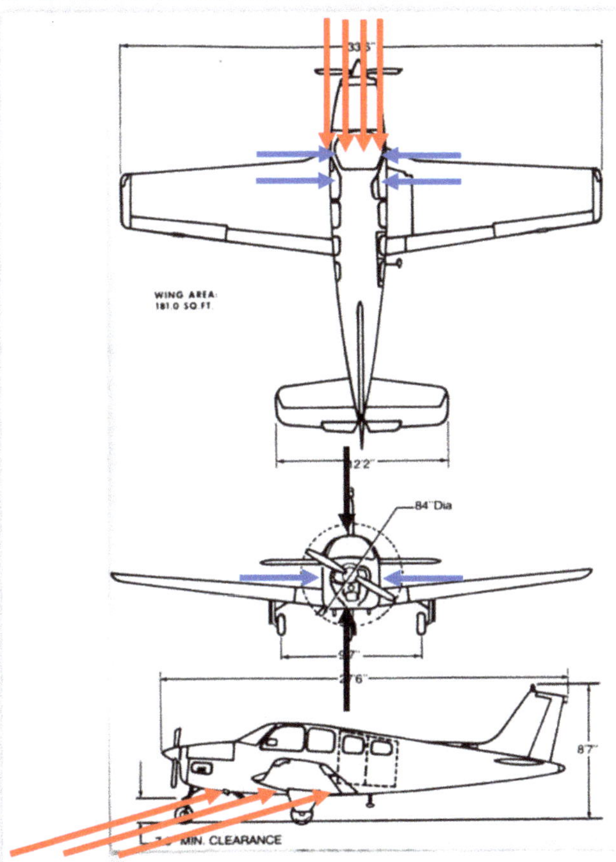

of an airframe would be deformed, such as on Figure 4.10, where the red arrows represent the direction of impact loading that resulted in upward and aft deformation of the fuselage front spar carry-through structure.

Using these impact force vectors also helps evaluate the *sequence* of deformation and failures, which is important when considering that overall crash damage usually occurs in just a fraction of a second. The sequence of deformation and impact marks on or between components in ductile/collapsible airframe structures typically starts closest to the initially impacted area, then works along the general direction of the impact force. For aircraft that impact terrain in a non-inverted attitude, the more beefed-up parts of the airframe structure, such as a well-designed cockpit/cabin, will typically maintain most of its volume and not start appreciably deforming until the forward fuselage and/or subfloor structure (and fixed or extended landing gear during ground impacts) has mostly crushed/collapsed.

The impact force vectors shown in Figures 4.9 and 4.10 resulted in the main cabin door separating from the airplane. The accident occurred during a low-level and high-speed flyby in very gusty wind conditions. The door was suspected to have popped open during the flight, which may have caused a loud rush of air to enter the cockpit that startled/distracted the pilot enough to momentarily allow the airplane to pitch nose-down and impact terrain. The door possibly being open during impact was a valid consideration based on the [i] right-wing walk area exhibiting a pronounced linear impact mark (Figure 4.11); [ii] cabin door perimeter and aft latch bolt being relatively

FIGURE 4.10 Forward side of exemplar A36 Bonanza front spar carry-through structure, and left side view of one that was deformed during terrain impact (red arrows show general direction of impact forces).

© CSOBeech

FIGURE 4.11 Crease found on right-wing walk area from impact with lower edge of main cabin door.

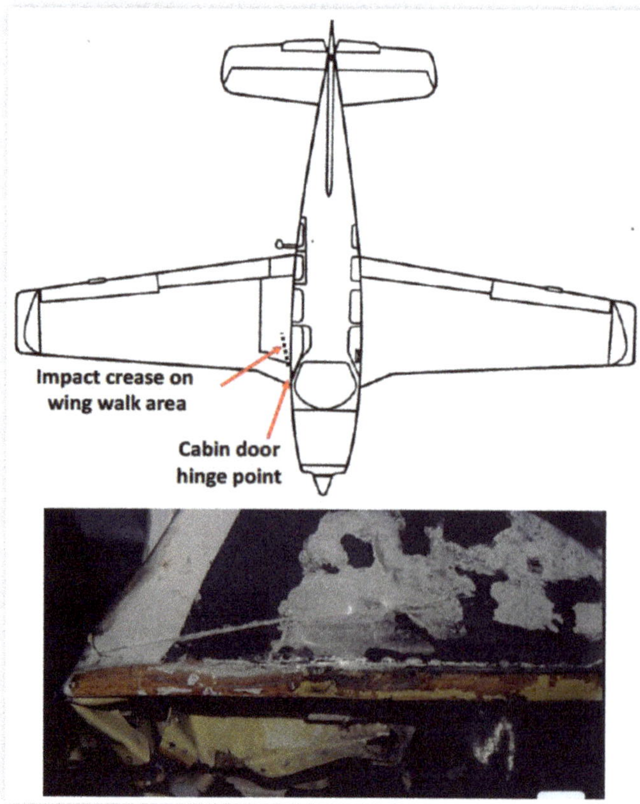

© Don Knutson

FIGURE 4.12 Non-deformed aft edge of main cabin door with protruding latch bolt, compared to its adjacent fuselage cabin door frame and bolt receptacle that exhibit notable deformation.

© Don Knutson

undamaged compared to its deformed fuselage door frame and latch bolt receptacle (Figure 4.12); and [iii] upper latch hook on the cabin door was not deformed or broken off, which is consistent with the latch hook not being engaged with its receptacle during airframe fragmentation (Figure 4.13).

However, additional investigation found that [i] the lower cabin door frame exhibited inward buckling (Figure 4.14) along with the cabin door stop exhibiting an indentation from being sandwiched between the cabin door and its door frame during terrain impact (Figure 4.15); [ii] an unlatched cabin door will consistently trail open approximately 3 in. at airspeeds that the subject airplane was flying (per flight tests performed on different exemplar airplanes), which would place the door stop well outboard of where the

FIGURE 4.13 Views of main cabin door upper latch hook moved to a fully engaged position – note that the hook and its fuselage door frame receptacle remain intact even though the airframe was fragmented.

© Don Knutson

FIGURE 4.14 Lower cabin door frame exhibiting inward displacement and buckling.

© Don Knutson

indentation on the lower door frame was found; and [iii] the push force (or deceleration) on the aft latch bolt required to unlock the door from a fully latched position was less than the longitudinal deceleration force imparted on the airplane during terrain impact (per engineering dynamics calculations, mechanical testing, and physical measurements of the overall cabin door latching mechanism [Figure 4.16]). These findings ultimately proved that the main cabin door was *closed* during the initial terrain impact.

Matching the overall aircraft deformation and component separations with the initial impact area or ground plane also helps determine the sequence of events during a crash. An example of this technique involved an airplane that crashed in a corn field as shown in Figure 4.17. At first glance, the resultant crash damage shown in Figure 4.18 could indicate that the airplane stalled and initially impacted the ground while in a nose-down and right-banked attitude. But when matching the ground plane with the airplane's crush lines (along the lower forward and right side of the fuselage cabin) as depicted in Figure 4.19, we end up with a three-dimensional view that is inconsistent with the lack of damage to other parts of the airplane, such as no right wing leading edge damage.

As we look closer at the wreckage, the engine had separated from its mounting downward and to the right during the impact sequence (Figure 4.20), which is opposite the direction of engine displacement we would expect during a nose-down and right-banked impact attitude (i.e., the engine was not displaced upward and to the left, nor did it impact the upper half of the firewall assembly). From these findings, we can look even deeper into the crash damage and sequence to find what caused the engine to separate from its mounts. In this case, an incomplete fishmouth weld attaching the castering nose wheel assembly (Figure 4.21) was found to be the weakest link, and determined to have broken away from the nose gear strut during initial ground impact (FEA and structural engineering calculations were used to analyze the impact loads). The exposed nose gear strut tube (Figure 4.22) impaled into the soil, then separated and

FIGURE 4.15 Views of indentation on lower portion of main cabin door frame due to the main cabin door stop (top view) being sandwiched between the door and frame assembly.

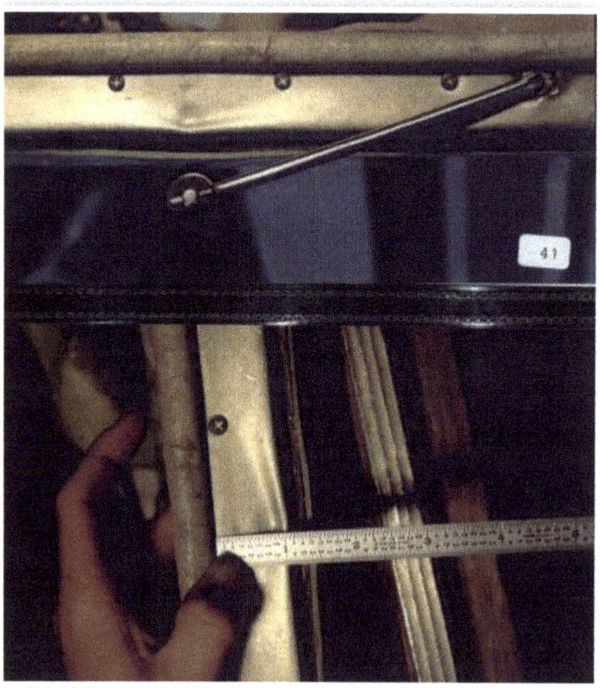

© Don Knutson

FIGURE 4.16 Main cabin door latching mechanism is interconnected and locks/unlocks at the same time when the inside door handle is positioned to close/open.

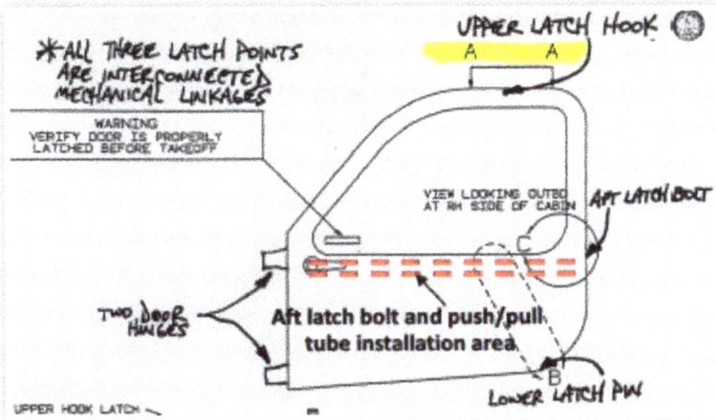

FIGURE 4.17 Accident site looking upstream from main wreckage (top view), and forward fuselage with engine and nose wheel separated (bottom view).

CHAPTER 4

FIGURE 4.18 Airplane three-view diagram describing and showing locations of primary crash damage.

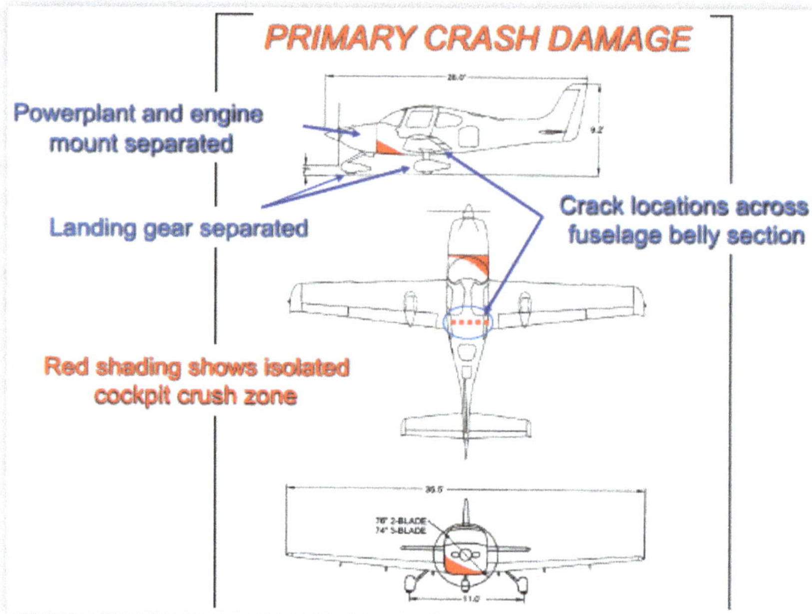

FIGURE 4.19 Three-dimensional depiction of an airframe oriented with the accident site terrain – in this case, the right wing would have sustained ground impact damage along with the engine being displaced upward and to the left.

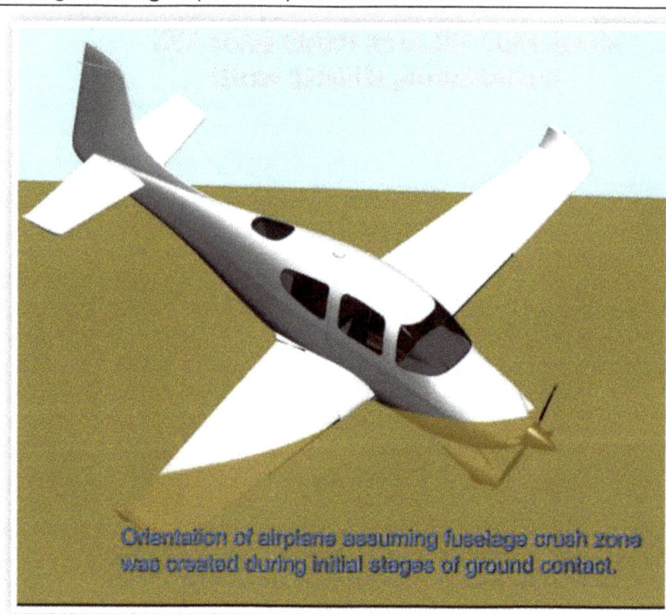

FIGURE 4.20 Views of right side of failed engine mount structure during terrain impact, which exhibits downward buckling.

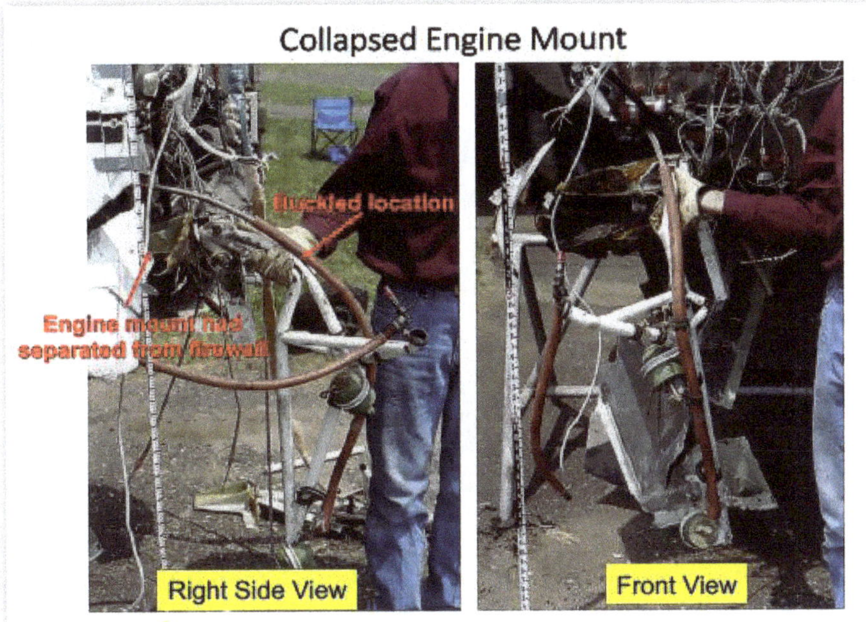

FIGURE 4.21 Views of nose landing gear that separated during terrain impact – wheel and fork assembly initially separated at fishmouth weld, which exposed the strut.

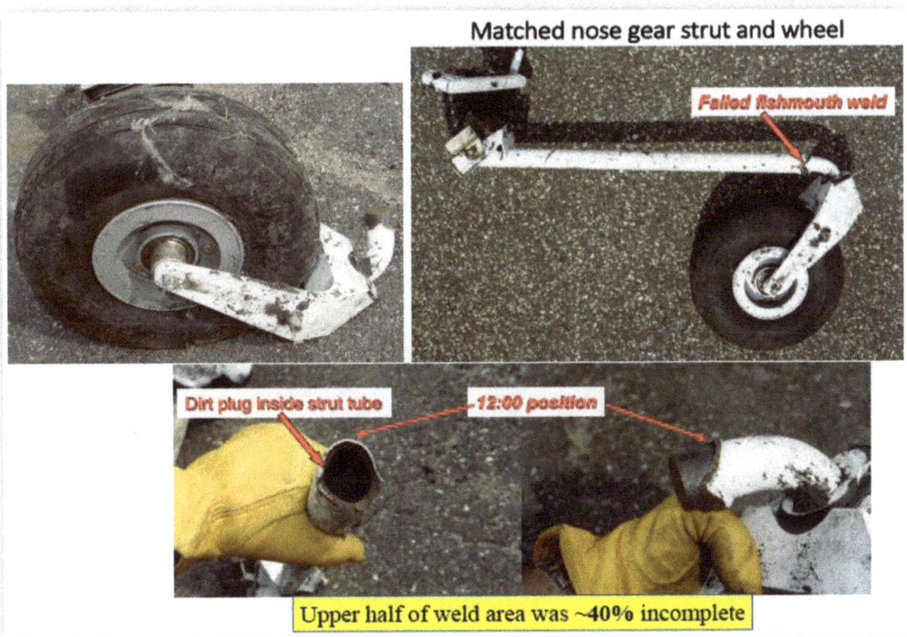

FIGURE 4.22 Nose landing gear after its wheel and fork assembly separated during initial terrain impact.

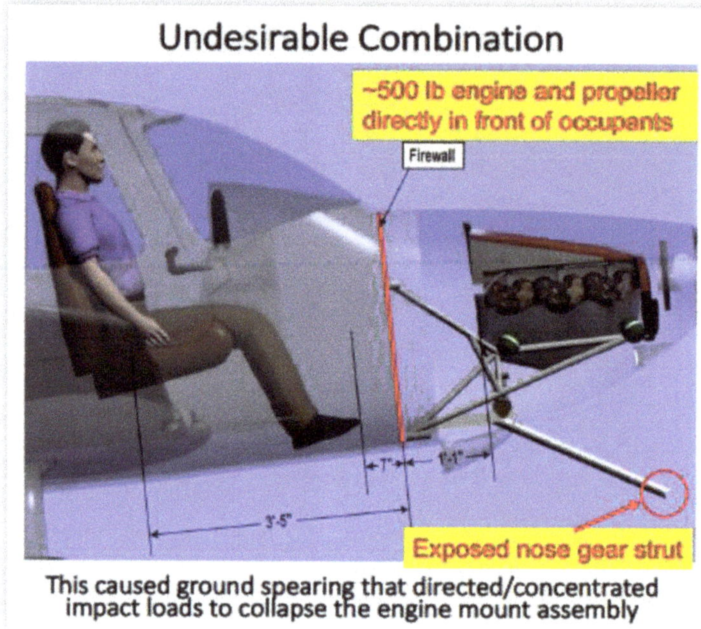

© Don Knutson

collided with the engine mount assembly (Figure 4.23). The engine mount collapsed causing the engine to separate and impact the oncoming lower fuselage cabin, which seriously injured the pilot and passenger.

Another example of evaluating the general pitch attitude of an aircraft can be seen when studying Figures 4.24 (an exemplar Bonanza A36 engine installation) and Figure 4.25 (Bonanza A36 firewall damage resulting from a high-speed crash). We can readily see that the engine did not impact the firewall on Figure 4.25. In this case, the engine was allowed to cleanly separate from its mounting during the initial terrain impact and travel ~500 ft before coming to rest. This means that the airplane's momentum

FIGURE 4.23 Three-dimensional layout views of nose landing gear and engine mount.

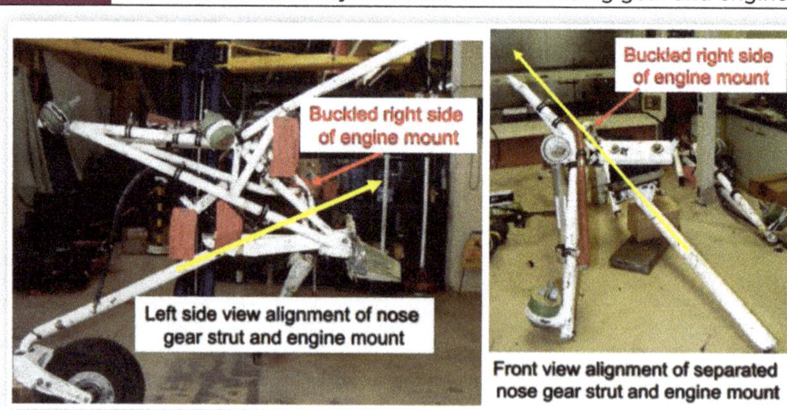

© Don Knutson

FIGURE 4.24 Exemplar Beech Bonanza engine installation, with cowling doors open - note that the aft part of engine and its accessories are located just inches forward of upper half of firewall.

© Don Knutson

FIGURE 4.25 Forward view of another Beech Bonanza engine firewall separated from its airframe during a high-speed impact - note how the upper half exhibits relatively minor crash deformation.

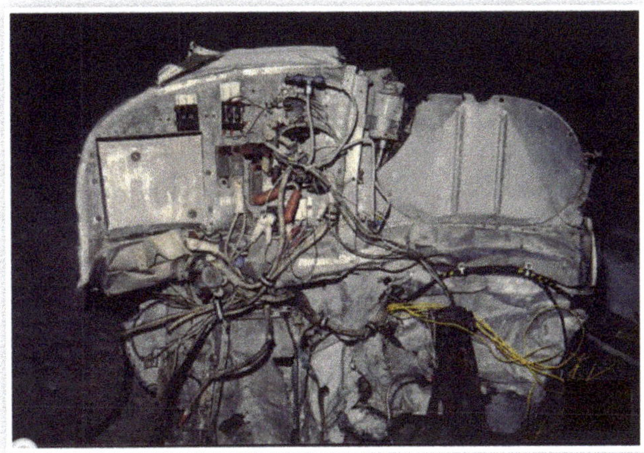

© Don Knutson

and pitch attitude did not allow the fuselage to plow into the engine, which is consistent with a nose level or positive pitch angle.

In summary of this section, the aforementioned wreckage damage and findings show good examples of how to consider the aircraft's attitude and interaction between its parts during a crash. Also, all the above diagrams and drawings help both *visualize* and evaluate crash damage, which are relatively simple to prepare and present. Report narratives can easily become difficult to decipher without using these types of visual aids.

Cable-Driven Flight Control Trim Tab Systems

As indicated in Chapter 3, these systems are most likely to maintain their pre-impact settings when the airframe remains relatively intact and its cables were not tensioned during the crash. When attempting to determine if the elevator, rudder, or aileron trim tab positions were at or close to their pre-impact positions, the following should be inspected and documented when applicable (an example of an elevator trim control system is shown on Figure 4.26, which is similar to the rudder or aileron trim tab control concepts):

- Measure the trim tab angle in relation to its control surface, and compare with its cockpit position indicator and actuator rod (jackscrew) length.
- Check the push/pull rod condition and length between the attachment bolt centerlines of the trim tab horn and actuator rod ends. The barrel of the push/pull rod may have witness marks where it transitions from inside to outside the primary control surface, which can correspond to a trim tab angle.
- Measure the length of chain on each side of the drive sprocket for both the cockpit trim control wheel and actuator itself - these measurements can be compared with an exemplar installation and corresponding trim tab setting (an example is shown on Figure 4.27). However, the accident aircraft might have a different deflection for a given measurement due to rigging differences.

FIGURE 4.26 Elevator trim tab control system on a Beech King Air 200.

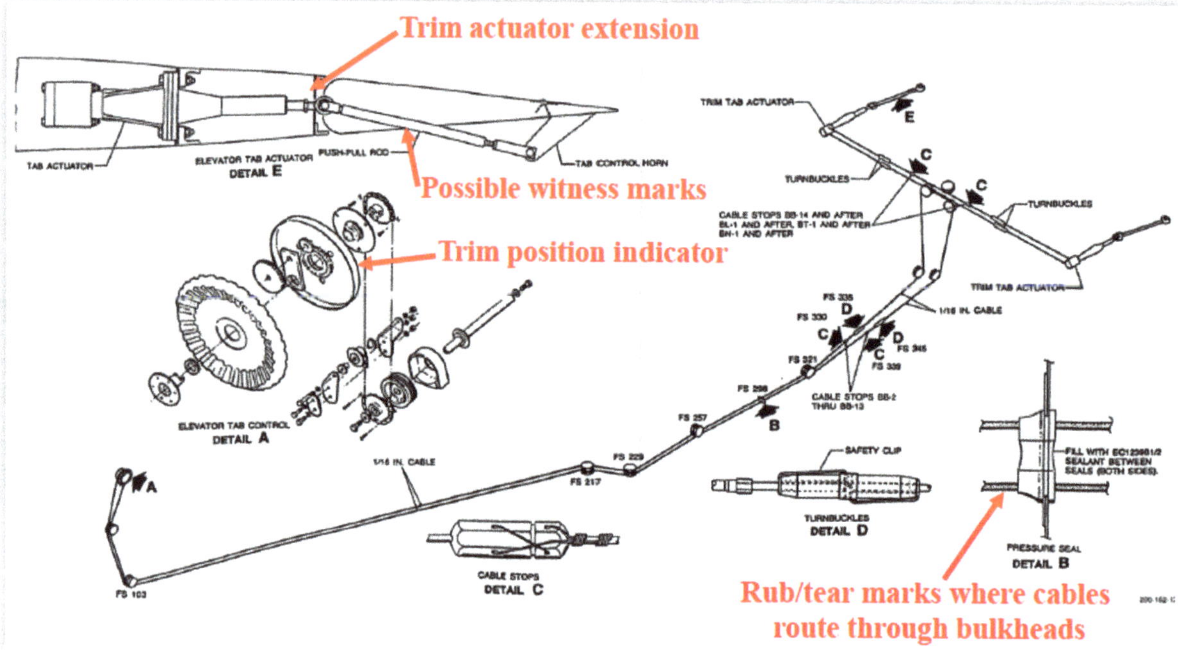

FIGURE 4.27 Cessna Citation aileron trim tab actuator chain and sprocket assembly mockup showing an over-traveled left roll setting (top view) and a neutral setting (bottom view), assuming proper rigging.

© Don Knutson

Electric Motor-Driven Actuators

Non-cable-driven controls such as wing flaps and retractable landing gear assemblies use these types of actuators. They involve motors that [i] rotate jackscrews either directly or by flexible rotary shafts (Figure 4.28) and [ii] drive gearbox and bell crank assemblies that translate push-pull tubes (Figures 4.29 and 4.30). If the actuator jackscrews, gears, or bell crank splines are not found stripped or fragmented, the wing flap and landing gear settings can be determined because these actuator settings will usually maintain the same or close to the same position as their pre-impact configurations.

FIGURE 4.28 Breakdown of a wing flap jackscrew actuator driven by a flexible rotary shaft (top view) and an actual actuator that is partially consumed by post-impact fire and exhibiting an extension of ~1.7 in. (bottom view).

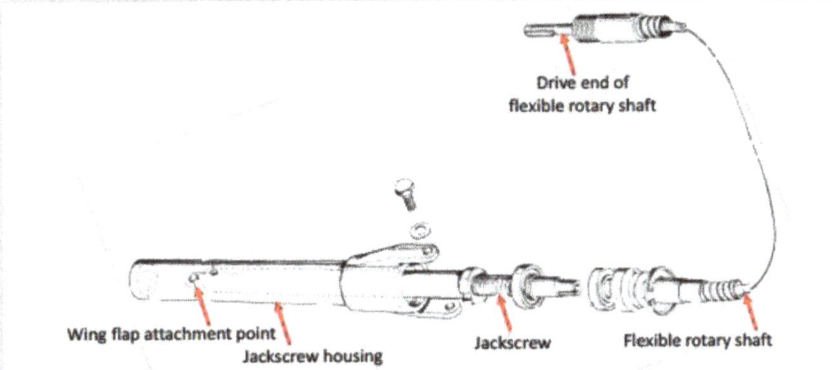

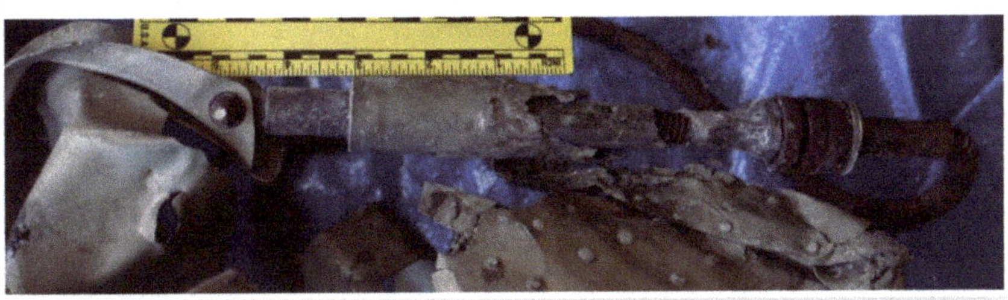

FIGURE 4.29 Disassembly of a landing gear actuator showing its gears, bell cranks, and electric motor.

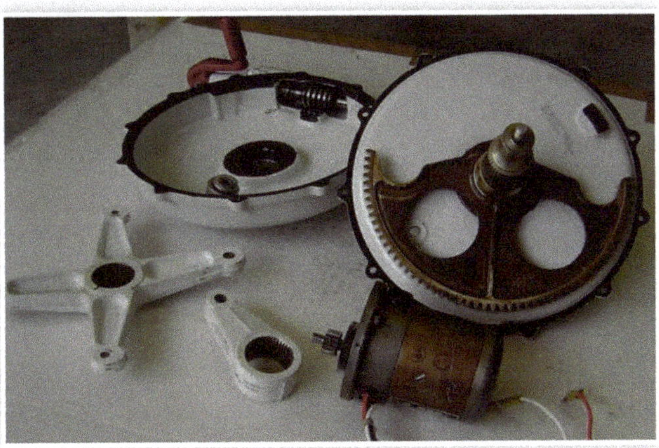

FIGURE 4.30 Landing gear actuator installation (similar to the disassembled actuator shown on Figure 4.29) that was involved in an accident, and shows its lower bell crank in a gear-extended position (and not loose from the bell crank being stripped on its splined shaft or an internally damaged gearbox).

Propeller Blade Impact Damage

Maximum propeller RPM is dictated by the need for blade tip tangential speeds to remain below the speed of sound.* A common range of maximum propeller RPM on aircraft is 2,000 to 2,700, depending on the propeller diameter. Propeller configurations can range from two to six blades; however, the basic principles for evaluating blade impact deformation does not change. Numerous factors go into understanding propeller blade impact deformation such as:

- Type of propeller control – fixed pitch, controllable pitch (change of pitch angle while propeller is rotating), constant speed (automatically changing pitch angle to maintain RPM during climb or descent), feathering (90° pitch angle during engine shutdown), and reverse pitch (negative pitch angle/thrust during landing)
- Material of the propeller blades (aluminum, wood, or composite)
- Amount of torque produced by the engine (e.g., high- or low-power settings)
- Rotational speed (RPM) of engine/propeller
- Relative impact speed and shape and hardness of objects that impact the propeller blade during a crash (e.g., landing into jagged/rocky terrain can result in deep leading edge gouging with or without the engine running)
- Propeller blade pitch angle during impact – within a normal operating range or feathered (Figure 4.31)
- Orientation of the aircraft and propeller's plane of rotation during the impact (pitch, bank, and yaw)
- Horizontal and vertical speed components of the aircraft during impact (e.g., an airplane in a wings-level attitude while flying with engine power at a slower speed and more gradual descent into terrain will more likely result in forward blade deformation than at a higher speed and rapid descent situation)

* This condition is critical because when the airflow over the cambered part of the blade tip's airfoil section approaches the speed of sound, a shock wave forms on the airfoil and creates a turbulent wake that substantially increases drag.

FIGURE 4.31 Propeller blade pitch angles.

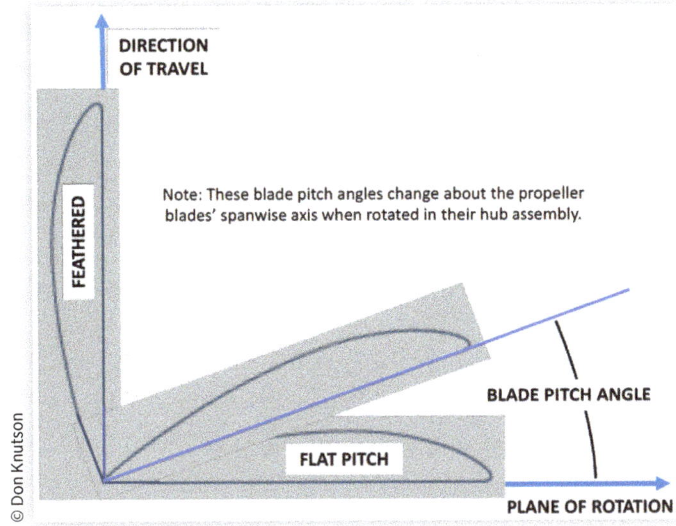

- Terrain and obstacles where the propeller blades impacted (soil, water, trees, buildings, asphalt runway surface, other aircraft, etc.)

We can generally characterize the engine power settings into high or low. High-power settings occur during the takeoff, climb, and cruise stages of flight. Low-power settings occur during a landing and taxiing before takeoff or after landing.

Rotational impact damage to wood and composite propeller blades (Figure 4.32) can be difficult to correlate with engine power settings due to their propensity to shatter at low *or* high-power conditions. These types of blades sometimes show evidence of a high engine-power setting when exhibiting similar fractured blade lengths. Trajectory analysis (covered in Chapter 6) can be utilized to estimate propeller RPM from a fragmented piece of blade that is launched a determined distance without obstacle interference.

Typical findings resulting from aluminum propeller blades impacting during a high-power setting and operating within a normal range of pitch angles include:

- Missing/sheared-off blade tips (Figure 4.33)
- Curled blade tips (aft or forward - depending on the propeller's plane of rotation angle relative to the terrain or object being cut through)

FIGURE 4.32 Front and back views of composite propeller blades that were fragmented when impacting terrain with the engine at a low-power setting (determined from other investigation findings).

CHAPTER 4

FIGURE 4.33 Aluminum propeller blade exhibiting missing tip and twisting toward low pitch.

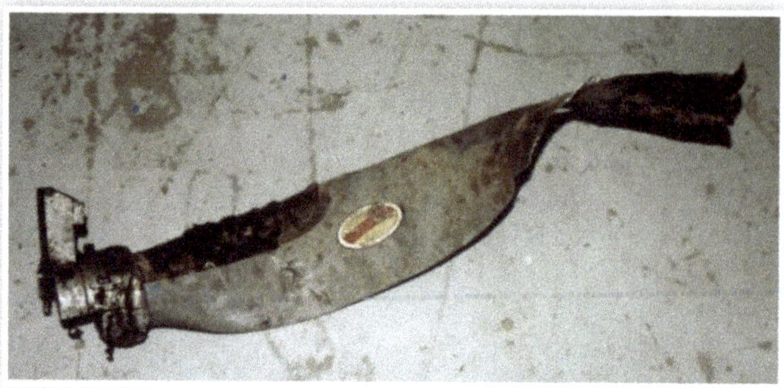

© National Transportation Safety Board

- Blade leading edge gouging and tearing
- Substantial chordwise gouging or scratching
- Pronounced blade twisting (beyond its design*) toward low pitch (Figure 4.33)
- Spanwise blade "S" bending (Figure 4.34)
- Tree limbs and branches exhibiting straight/clean cuts (sometimes with paint transfer from propeller blade)
- Propeller assembly (hub and blades) longitudinally stripped away from its bolt/stud threads and engine mounting flange while impacting water in a positive/nose-up attitude (Figure 4.35) – this is due to the propeller blades being free to continue rotating and moving forward through the water medium while the airplane is decelerating

Typical findings resulting from aluminum propeller blades impacting during a low-power setting include:

- No or relatively minor blade bending/damage
- A lesser degree of blade twisting or chordwise scratching (see note)
- Blade exhibiting a relatively sharp aft bend

FIGURE 4.34 Aluminum propeller blade exhibiting "S" bending.

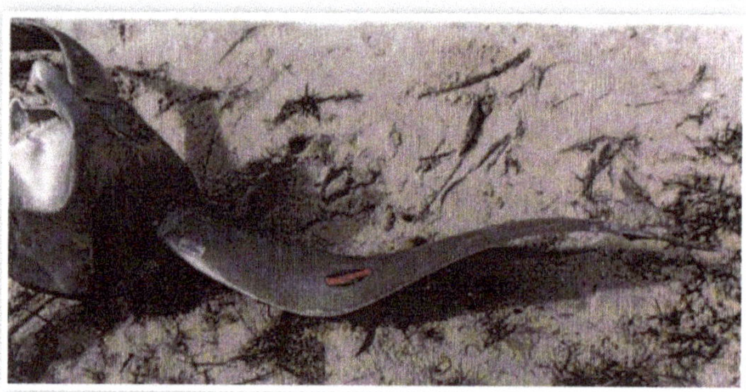

© Don Knutson

* Propeller blades are designed with a twist to keep thrust equal along their span; therefore, the blade is thicker at the hub with a larger blade angle and progressively gets thinner toward the tip with a lower blade angle.

Note: Be careful with concluding that the engine was producing power due to the blades exhibiting some twisting, chordwise scratching, or leading edge polishing and gouging - this is because a windmilling propeller during a high-speed descent can create enough rotational energy to result in impact deformation similar to the engine producing some power. Also, a stationary (perhaps feathered) blade can be dragged leading edge first through dirt, rocks, buildings, etc.

Typical findings resulting from propeller blades impacting while in a feathered (90° blade-pitch angle) position include:

- Blades remaining in or near a feathered position
- No blade twisting
- Some leading edge polishing
- Blades with shallow bends in opposite directions

Propeller blade slash marks found at the accident scene (e.g., on a runway (Figure 4.36) or emergency landing area) can help estimate the aircraft's velocity and propeller RPM setting during ground contact by using the following equations (depending on whether you wish to find MPH or knots):

- **RPM** = $\mathbf{V_{MPH}} \times \mathbf{88} / \mathbf{D_{sm}} \times \mathbf{N_b}$, where $\mathbf{V_{MPH}}$ is aircraft velocity in MPH, $\mathbf{D_{sm}}$ = distance between slash marks, and $\mathbf{N_b}$ = number of propeller blades
- **RPM** = $\mathbf{V_{kts}} \times \mathbf{101.3} / \mathbf{D_{sm}} \times \mathbf{N_b}$, where $\mathbf{V_{kts}}$ = aircraft velocity in knots

A propeller teardown inspection is required to determine the condition of the pitch change mechanism and specific blade angles at the time of impact. Be very careful during the propeller hub disassembly due to possibly strong compressed springs.

Turbine Engine Damage

The simplified sequence of how a turbine engine operates is suck, squeeze, burn, turn, and blow. Turbine engines (Figures 4.37 and 4.38) used on turboprop airplanes, helicopters, and jets have high-speed rotating components in their compressor and turbine sections*, which operate in a range of thousands to tens of thousands of RPM, depending on their diameters (or blade tip speeds remaining subsonic).

Factors that dictate the overall amount of turbine (or jet) engine damage include the aircraft velocity, engine impact angle in relation

FIGURE 4.35 Propeller mounting flange and attachment hardware (on forward end of engine crankshaft) where propeller assembly longitudinally stripped away from the stud threads.

© Don Knutson

FIGURE 4.36 Propeller strike marks on a paved runway surface (Beech King Air during landing with one main gear collapsed).

© Don Knutson

* Examples of turbine engine operating speeds are the Rolls-Royce 250-series (e.g., on a Bell 206 helicopter) running at over 50,000 RPM; Garrett TPE331 (e.g., on a Mitsubishi MU-2 turboprop) running at over 40,000 RPM; Pratt & Whitney PT-6A (e.g., on a Beechcraft King Air turboprop) running at over 37,000 RPM; Pratt & Whitney JT15D (e.g., on a Cessna Citation business jet) running at over 30,000 RPM; and CFM International CFM56 (e.g., on a Boeing 737 jet airliner) running up to ~15,000 RPM.

FIGURE 4.37 Diagram of a turboprop engine.

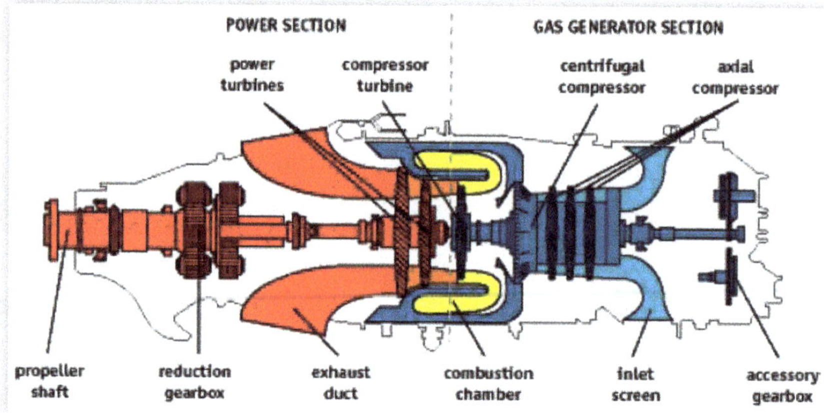

FIGURE 4.38 Simplified diagram of a turbofan engine with a split-spool compressor - orange shows connection between low-pressure compressor and aft two turbine wheels, and red shows connection between high-pressure compressor and forward turbine wheel.

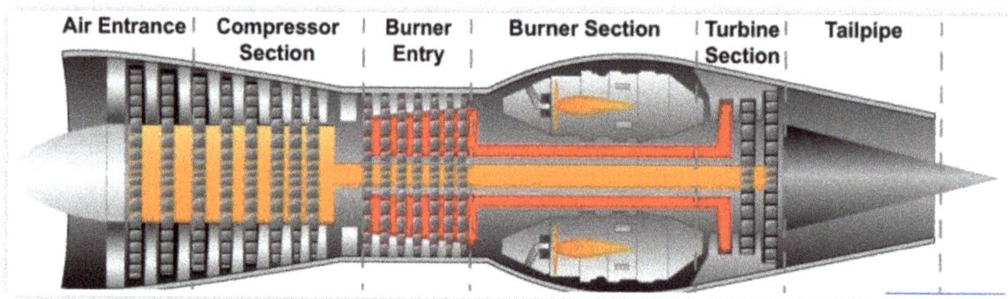

to the terrain, engine compressor/turbine RPM, crush damage to the engine case, and the condition and composition of the terrain impacted (hard, soft, loose debris to ingest, etc.). Similar to evaluating propeller blade damage, turbine engine compressor and turbine wheel blade damage can provide indications of the engine's power (RPM) settings during terrain impact. Here are some typical results of turbine engine damage during the following general impact and RPM conditions:

High-impact angle (engine's longitudinal axis in relation to the terrain) and high RPM

- Compressor turbine blades bent and broken opposite their direction of rotation
- Little or no ingestion (of foreign material) due to stoppage of rotating components from rapid crushing of engine case
- Uniform bending of power turbine blades (downstream of the compressor section)
- Shaft connecting turbine section to compressor section twisted or sheared if compressor seized

High impact angle and low RPM

- Nonuniform or localized compressor blade bending (bent both with and opposite their direction of rotation)
- Little or no ingestion
- Nonuniform or localized bending of power turbine blades

Low impact angle and high RPM

- Compressor blade wipeout (all or most blades broken off due to ingestion)
- Heavy ingestion throughout engine, especially charred debris in turbine section
- Uniform bending and possible fractures of power turbine blades
- Shaft connecting turbine section to compressor section twisted or sheared if compressor seized
- "Machined" and melted material (from rotating parts rubbing together)
- Molten metal spray on the turbine nozzle vanes or turbine wheel blades (Figure 4.39)

Low impact angle and low RPM

- Nonuniform or localized compressor blade bending
- No appreciable ingestion downstream of compressor blades
- Nonuniform damage to power turbine blades
- Little or no damage to power turbine shaft

Low impact angle with no significant RPM (Figure 4.40)

- Only a portion of the compressor blades is bent or broken away that result from crushing of the compressor case.
- No ingestion other than possible debris in first few stages of compressor.
- Axial compressor and turbine blade tips and centrifugal compressor edges create impact imprints on their adjacent shrouded surface.

FIGURE 4.39 Metal specks adhered to turbine blades from molten metal spray during operation.

FIGURE 4.40 Turbofan engine inlet vertically crushed with no significant engine RPM.

Keep in mind that turbine engines running at a high RPM that remain elevated off the terrain during a crash can result in little to no ingestion into the compressor section, due to high-mounted engines (such as on helicopters) or lateral/wing-first airplane impacts where the engine does not contact the terrain.

Cockpit Instrumentation

Electronic (digital) or "glass cockpit" instruments (e.g., GPS, EGPWS, and flight management systems with multifunction displays) contain memory chips that store flight data and systems information that can be downloaded, providing they were not destroyed in the crash. If the memory chips are destroyed, then we are left with evaluating the available *mechanical* failures.

Switches in the cockpit where the handle or lever needs to be pulled out/aft/sideways to change settings (Figure 4.41) can typically hold their positions during a crash when the cockpit area was not fragmented, and negligible damage occurred to their detents and immediate surrounding components. Verifying positions of the components that the switches control (e.g., wing flaps, landing gear, control valves, and light bulb filament

FIGURE 4.41 Landing gear, wing flaps, and electrical switches that require to be pulled out/aft/sideways to change their settings.

© Purdue Aviation Education Multimedia Library

damage) can also give us an idea whether the switches moved during impact. Also, detents or gates that are worn (versus damaged) might allow inadvertent movement or not hold the switch positions during impact.

Analog and dial-type (mechanical) instruments such as navigation/communication (NAV/COMM) radios and engine tachometer and Hobbs meter hour readings (Figure 4.42) typically hold their pre-impact settings after a crash providing that the instruments did not fragment or sustain appreciable damage to their indicator face. Altimeter settings (inches of mercury reading in the Kollsman window*) typically hold

FIGURE 4.42 NAV/COMM radio frequency settings, and Hobbs meter and engine tachometer readings 3,630.7 and 0.0 h, respectively.

© Don Knutson

* The Kollsman window is a small window in an aircraft altimeter used to calibrate/adjust the instrument and compensate for atmospheric pressure changes. It was named after Mr. Paul Kollsman, who was a German-American inventor of barometers and instruments for instrument flight in airplanes.

close to or at their pre-impact settings even when the instrument is fragmented and just its face is intact (Figure 4.43).

FIGURE 4.43 Altimeter face, with 30.06 in. of mercury setting in Kollsman window.

© Don Knutson

Analog gauge needle (slap) marks may be created during impacts with high deceleration/g-loads which are approximately parallel to the aircraft's longitudinal axis. The tip and mid-section of the indicator's needle is usually what makes contact with the face of the gauge (Figure 4.44). Keep in mind that the mass of the needle is very low and its contact with the face needs to be hard enough to create an imprint; therefore, the aircraft's longitudinal impact loads need to be at least in the *hundreds* of G's. Sometimes a frontal/longitudinal impact with an object inside the aircraft can sandwich the needle between its glass cover and face, which usually results in wiping/scraping of the needle against its face. In this case, the beginning of the scrape marks can be considered as to where the needle was approximately positioned during impact.

Needle slap marks that are below the indicator's minimum scale (Figure 4.45) should be compared to an exemplar instrument installation to verify that the needle can actually position itself to that off-scale indication. Attitude indicators and directional heading gyro instruments with

FIGURE 4.44 Faint airspeed and altimeter needle slap marks (two parallel lines) on their indicator face (located with arrows).

© National Transportation Safety Board

FIGURE 4.45 Views of needle slap mark on face of breathing oxygen pressure gauge (offset to the left of zero).*

© National Transportation Safety Board

* Obtained from NTSB Docket DCA00MA005, Learjet 35, N47BA, accident at Aberdeen, SD, on October 25, 1999, loss of cabin pressurization (involved professional golfer Payne Stewart).

FIGURE 4.46 Exemplar attitude indicator compared with one involved in a high-speed impact (top view) – note the arrows pointing to the "OFF" warning flag slap mark (bottom view).*

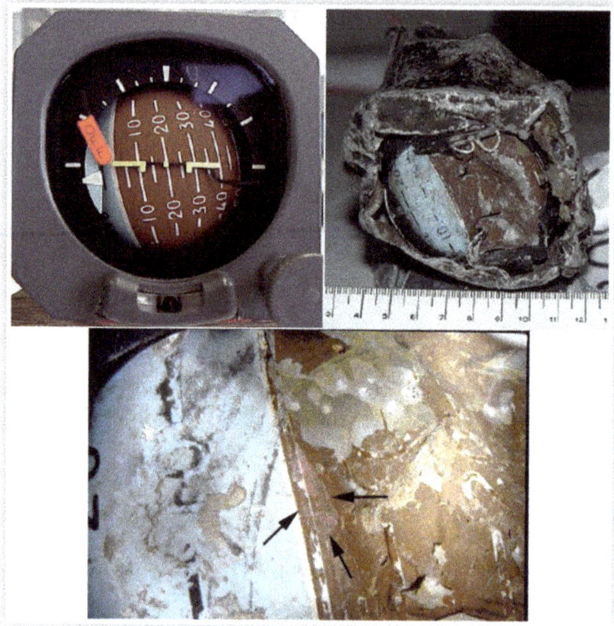

FIGURE 4.47 Landing gear position/warning lights with a GE-327 light bulb removed (this type of light bulb is commonly used in GA aircraft).

bright colored warning flags displayed in front of the indicator's face when the instrument is inoperative can leave a distinctive impact and paint transfer mark (Figure 4.46).

Tungsten filament deformation (or stretching) on cockpit warning/caution/position lights or aircraft position and landing lights can help us determine flight conditions and whether systems were activated during a high-speed crash (Figure 4.47, Figure 4.48, and

* Transportation Safety Board of Canada, Report Number A98H0003, In-Flight Fire Leading to Collision with Water, Swissair Transport Limited, McDonnell Douglas MD-11, HB-IWF, Peggy's Cove, Nova Scotia 5 nm SW, September 2, 1998.

FIGURE 4.48 Illuminated GE-327 light bulb with their filaments subjected to ~300 G's for 1 millisecond (deceleration forces were applied along the light bulb's longitudinal axis).*

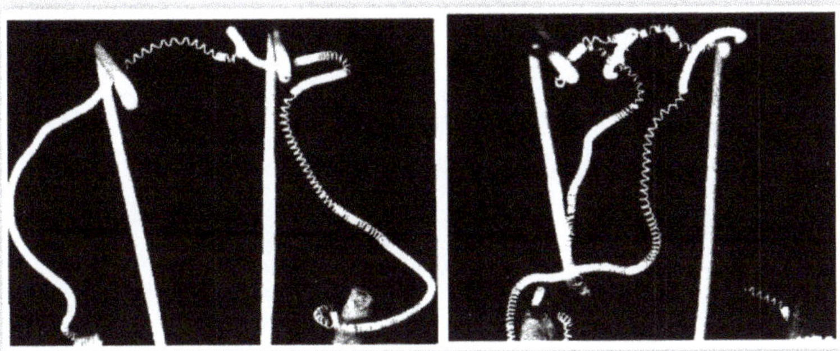

FIGURE 4.49 Wingtip position light bulb filament that was stretched/illuminated during impact.†

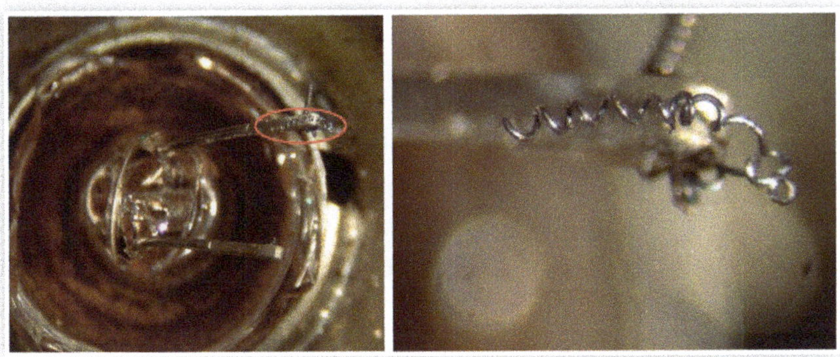

Figure 4.49. A good reference is *A Guide to Light Bulb Analysis In Support of Aircraft Accident Investigation*, Transportation Safety Board of Canada, 1985. Tungsten is usually much more ductile when heated and less ductile (typically less than 10% elongation before fracture) when cold, so a filament found notably stretched most likely indicates that the light bulb was illuminated during impact loading. Conversely, a filament found fractured and not appreciably stretched most likely indicates that the light bulb was not illuminated. An illuminated filament's ductility tends to increase with its age/usage (Figure 4.50). Impact testing has also shown that a new filament with no heat cycles can somewhat stretch before breaking. A metallurgical laboratory is typically needed for comprehensive evaluation of light bulb damage.

Air (or vacuum-driven) attitude indicators (via engine-driven air pumps) are commonly examined when a loss of flight control is suspected. When the gyro rotor spins at over 10,000 RPM in close tolerance (hundredths of an inch) to its housing (or capsule) during instrument fragmentation or deformation, the result can be substantial rotational contact between the rotor and capsule. When no notable rotation signatures are found, the attitude indicator most likely was not operating pre-impact. Keep in mind that evidence of rotation is not necessarily evidence of *proper* operation because the rotor RPM may

* *A Guide to Light Bulb Analysis in Support of Aircraft Accident Investigation*, Transportation Safety Board of Canada, 1985.

† Position or navigation lights give other aircraft information on the subject aircraft's position and heading by locating a red light on the left wingtip, a green light on the right wingtip, and a white light on the aft end of the fuselage or tail cone.

FIGURE 4.50 Filament stretching depending on its usage.

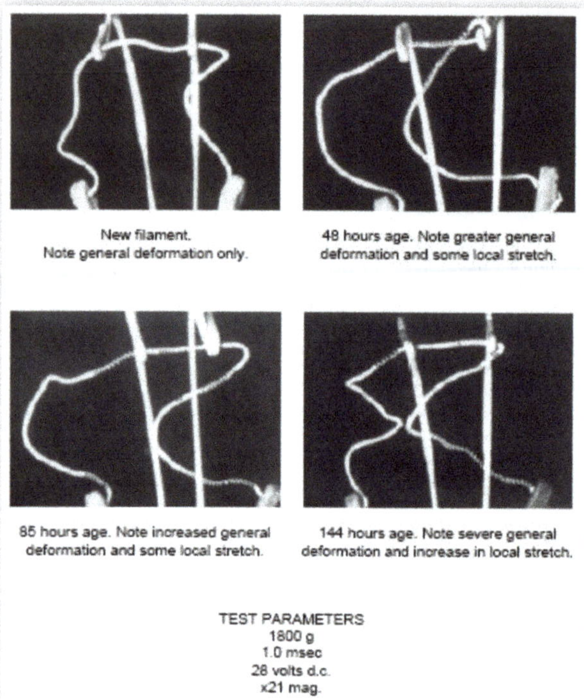

FIGURE 4.51 Fragmented vacuum-driven attitude indicator with its gyro rotor and capsule (showing nonlinear gouging on the rotor with no rotational impact marks inside or on the capsule).

have been too low. Also, a fragmented/crushed attitude indicator that exhibits nonlinear gouging on just a portion of the rotor circumference along with no appreciable rotational or impact marks inside or on the capsule is consistent with the rotor not (or not appreciably) spinning during the crash (Figure 4.51).

Who Was Flying the Aircraft?

When two pilots are seated in the cockpit, it is important to try to determine who was flying the airplane at the time of impact. Likewise, when incapacitation is suspected in a single-pilot aircraft, it is good to know whether he/she was in control at the time of the crash. Three typical areas of interest in the wreckage are the condition of the cockpit flight control yoke or stick, deformation of the powerplant control levers (throttle, propeller RPM, and mixture), and the front seat positions/adjustments. In addition, autopsy findings and medical records need to be researched. Knowing the person's height and pants inseam can help determine whether their feet could reach the rudder pedals or their eye level in relation to the windshield was adjusted properly (based on the front seat positions/adjustments). It is also helpful to understand the general direction of how the person[s] would flail/move in the cockpit during the crash.

The pilot grasping the control yoke or stick during terrain impact can result in joint dislocations, bone fractures, and bruising in the area between the thumb and index finger. This can also result in fracture or dislocation of the wrist. Various bone fractures in the pilot's feet can occur depending on where the pilot's foot was pushing against the rudder pedal during the crash. Imprints of the rudder pedal's surface pattern or shape found on

FIGURE 4.52 Two views of cockpit powerplant controls showing the right throttle lever bent and twisted counterclockwise (note: levers were originally found in a full forward position at accident site).

the pilot's hard-soled shoes or boots could also be evidence of pushing on the rudder pedal. A hand manipulating the engine controls (lever or vernier type) during a crash (e.g., throttle or mixture) can result in bruising, abrasions, or lacerations on the palm and fingers.

As with all specialized subject matter, we need to make sure the findings are supported by the appropriate expertise or reputable/reliable documentation. In this case, biomechanics and pathology experts or data can be utilized to evaluate the cause of injuries to the pilot[s].

A good example of this investigative technique is a twin-engine airplane that crashed with two very qualified pilots in the left and right front seats, and the investigators wanted to determine who was flying the airplane during the crash landing. The wreckage examination revealed that the entire cabin nose section exhibited primarily longitudinal/aft crushing, with more extensive structural tearing and separation noted on its left side. This was consistent with the occupants flailing forward and to the left during the ground impact. The right and left engine throttle levers (in the cockpit) exhibited bending to the left (Figure 4.52). The right engine throttle lever also exhibited counterclockwise twisting with no apparent impact marks on its knob. The autopsy report on the pilot in the left front seat noted his cause of death to be from blunt force, which included a laceration on his left hand and no significant injuries to his right hand. The pilot in the right front seat survived, and medical records showed injuries to his left hand, including a laceration on the index finger. Biomechanics analysis was also performed, which showed that the right throttle lever deformation would be highly improbable if the pilot in the left seat was grasping the knob with his right hand during the crash. Considering all the above-noted findings, the pilot in the right front seat was most likely grasping the right throttle lever with his left hand during the crash.

Basic Failure Analysis of Metal

In addition to mapping and describing the direction, orientation, and amount of impact damage and deformation, we need to establish why materials fracture. Metals are overstressed when the structural load exceeds their elastic range, and break apart when exceeding their ultimate strength.* In addition to the metal being pulled, compressed, bended, or twisted, other factors can contribute to its failure such as cycled/reversed loading (fatigue), permanent deformation from long-term exposure to high levels of

* Before reaching ultimate strength/failure, the material transitions through its elastic (springback) and plastic (permanent set) deformation phases.

stress below the material's yield strength* (creep), corrosion effects (e.g., from oxidation, dissimilar metal contact, and improper heat treatments), and temperature transitions (e.g., metal becoming brittle at very cold temperatures while being subjected to a load).

Aluminum and steel alloy overloads are commonly characterized by tensile versus compression, ductile† versus brittle‡, and instantaneous versus progressive (or fatigue). Laboratory examination and sectioning of the damaged materials will be necessary when determining further details of why the fractures ultimately occurred, such as their composition/alloys, mechanical strength, or corrosion resistance.

Even if you do not have a "metallurgist" title, you should still be able to identify general failure modes. A vast number of published examples/photos showing various metal failures can easily be found in books and online.§ Some of the more frequent failures encountered during wreckage reconstruction along with the typical results of those failures are:

Ductile metal, with *tensile* overload (from a pulling or bending action)

- Elongation and/or bending deformation prior to fracture
- "Cup and cone" fracture surface
- Dull, fibrous fracture surface

Ductile metal, with *torsional* overload

- Twisting deformation prior to fracture
- Roughly square (~90°) fracture profile

Brittle metal, with *tensile* overload

- Very little deformation
- Roughly square fracture profile
- Chevrons ("V" shaped marks) that point back toward fracture origin

Brittle metal, with *torsional* overload

- Very little deformation
- Diagonal (~45°) fracture profile

Metal *fatigue*¶ from cycled or oscillated loading

- "Beach" marks – radius of curvature increases as crack progresses (Figures 4.53 and 4.54).

---

* A point where the stress on the material normally transitions from elastic to plastic deformation.

† Ductile materials can be bent and twisted with no cracking and tend to hold their deformation that occurs in the plastic region (past their elastic range). Common ductile aircraft materials include wrought aluminum, copper, and lower strength steels.

‡ Brittle materials fracture without any significant plastic deformation. Common brittle aircraft materials include cast aluminum, cast magnesium, and high-strength steels.

§ Some useful and practical references for metal and composite failures are the ICAO Document 9756, Manual of Aircraft Accident and Incident Investigation, Part III, Chapter 9, Structures Investigation, ISBN 9789292319335; Fatigue and Fracture - Understanding the Basics, ISBN 9781615039760 (2012); and Failure Analysis of Engineering Structures, Methodology and Case Histories, ISBN: 9780871708205 (2005).

¶ An informative article is in ASM International's *Journal of Failure Analysis and Prevention*, Volume 5(2) April 2005, pages 11-15, "Understanding the Surface Features of Fatigue Fractures: How They Describe the Failure Cause and the Failure History," written by Neville W. Sachs, P.E.

FIGURE 4.53 "Beach" marks from fatigue cracking on a fractured piston engine crankshaft with no needed magnification - notice how the striations fan out from the initiation site on top part of fracture.

© Don Knutson

FIGURE 4.54 "Beach" marks consistent with fatigue cracking on an outboard section of a fractured turbine blade (top view), including a 500X magnification showing the initiation site (bottom view).*

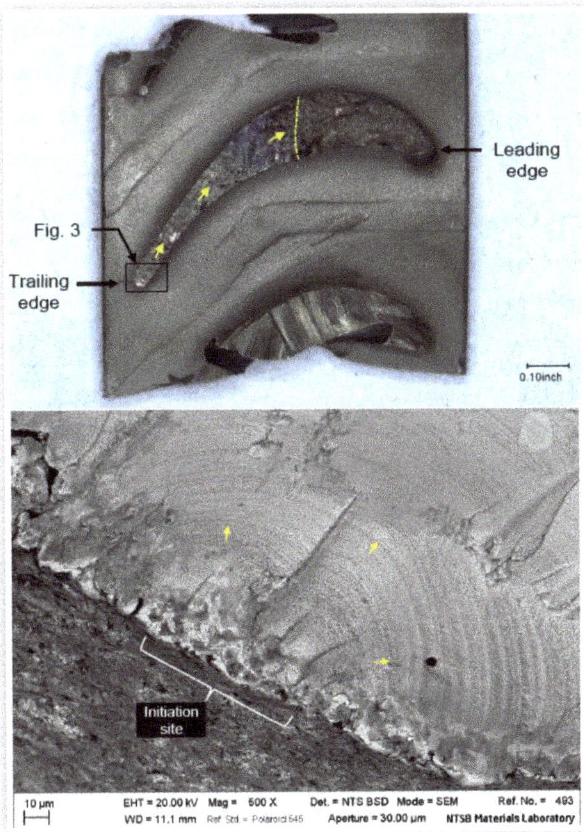

© National Transportation Safety Board

* Obtained from NTSB Docket WPR13GA374.

FIGURE 4.55 Flight control cables showing tension overload.

© Don Knutson

FIGURE 4.56 Fiber pullout of a composite matrix due to tension overload.

© Don Knutson

FIGURE 4.57 Tension breaks on composite wing lower spar cap (from a glider in-flight breakup).

© Don Knutson

- "Ratchet" marks (small offset steps) from more than one crack origin.
- Cracks commonly originate at a stress concentration caused by a sharp radius, surface damage such as tooling marks or loose assemblies allowing materials to flex beyond their designed limits.
- Two-zone fracture – fatigue and overload signatures found on the same fracture surface.

Metal corrosion

- Pitted surface
- Blistered/bubbled or flakey surface
- Whitish powdery area on aluminum and magnesium surfaces
- Reddish-brown oxide (rust) on steel surfaces

Flight control cables, with tensile overload (Figure 4.55)

- Broken ends are unraveled and frayed.
- Individual strands "necked down"/stretched.
- Portion of cable in proximity of break is coiled due to being extensively stretched.

Basic Composite Failure Analysis

Aircraft composites are usually made of carbon or glass fibers that are held together by a matrix, which is a glue where the fibers are imbedded. Fibers are assembled in plies and layered in different directions from 0° to 90° (oriented from longitudinal axis) within the matrix to give the composite specified directional load-carrying capabilities (e.g., 0° plies supporting axial loads and ±45° plies supporting shear loads). A good basic reference for advanced composite materials is in the FAA *Aviation Maintenance Technician (AMT) Handbook – Airframe*, Volume 1, Chapter 7.

Composite airframes do not behave the same as metal. Composites commonly will shatter when fractured, which can make it difficult to determine progressive or preexisting failure. Evaluating composite structural damage tends to be more focused at the initiating point of failure due to its brittle/rigid material properties. Multidirectional ply arrangement in the matrix can cause composite structures to behave differently to loading when compared to metal. For example, tensile loads can cause a composite to twist, and twisting loads can cause a composite to bend. Also, even though the composite is loaded in tension or compression, fiber orientation or imperfections within the material (e.g., voids in matrix or poor adhesion between plies) can result in different overall failure signatures.

FIGURE 4.58 Compression failures on composite wing upper spar cap (from a glider in-flight breakup).

FIGURE 4.59 Chop marks on composite fibers from compressive loading.

Overloaded fibers within the composite will usually behave in a consistent manner. Tension breaks commonly exhibit "fiber pullout" or unkinked fibers that remain relatively parallel to and along the direction of the load (Figures 4.56 and 4.57). Compression breaks exhibit buckled/kinked fibers relatively local to the fractured area (Figure 4.58). Another common characteristic of compression failure is evidence of "chop marks," which exhibits fractured fibers with a relatively smooth and rough surface separated by a neutral/bending axis – the smooth side is under compression and the rough side under tensile load (Figure 4.59).

Impact loads typically cause delamination or splits between the composite plies and can be hidden from the inspector. Tapping a coin along the surface and listening for a relatively "deadened" or dull sound can sometimes identify delamination. A more defined approach is by cross-sectioning the composite at its delaminated location or by using NDT such as ultrasonic* or X-ray tomography.†

---

* Ultrasonic NDT uses a transmitting/receiving transducer on the accessible side of the part, which detects reflections of sound from within the part and electronically displays the responses on an oscilloscope to interpret discontinuities.

† X-ray tomography is a method for generating three-dimensional imaged volumes from two-dimensional X-ray image slices. The X-ray imaging is based on the differential absorption or scattering of an X-ray source to reveal the internal attributes of a part.

CHAPTER 5

Crashworthiness Analysis

Designing aircraft for crash safety is not as easy to achieve when compared to many other means of transportation because aircraft need to contend more with the Z-axis (altitude). Also, helpful aircraft crash testing/research has been performed but is very limited when compared to the automotive industry. This chapter gives the investigator some practical focus on [i] when it is reasonable to perform crashworthiness analysis, [ii] what to inspect when evaluating how the aircraft could be more crashworthy, and [iii] how to estimate crash loads experienced by the occupants.

Implementing Crashworthiness Investigations

Field investigators do not usually specialize in crashworthiness analysis, and this type of investigation on GA airplanes typically becomes an afterthought when realizing that the pilot and/or passengers should have been able to survive the crash or sustain lesser injuries. With this in mind, investigators should be ready to identify select circumstances that warrant specific data gathering to address possible crashworthy issues where resulting serious and fatal injuries were not expected. This happens when the cockpit/cabin volume remained intact and the aircraft impacted terrain at a relatively low speed, or post-impact fire created emergency egress problems.

Furthermore, the following investigation strategies can be used by aircraft (and related product) manufacturers to consider making crashworthy improvements to their designs:

1. Train investigators how to perform crashworthiness analysis.
2. Develop a report format that specifically targets/includes crashworthiness issues.

3. Use investigation findings to establish the aircraft impact speeds, attitudes, and specific structural/system damage that result (or do not result) in specific occupant injuries for each aircraft model.
4. Review accident investigation files involving known survivable accidents.
   a. Suspected minor injury accidents and circumstances that resulted in serious injuries - identify injury mechanisms
   b. Survivable accidents that resulted in fatal injuries - identify injury mechanisms
   c. Terrain impact conditions and aircraft configurations that resulted in survivable impacts
   d. Lawsuits based upon crashworthiness issues, which identify liabilities

Aircraft Design Expectations

Some helpful references for this subject are "Aircraft Crash Survival Design Guide," Volumes I through V, US Army, USAAVSCOM TR 89-D-22B (1989); and "Small Airplane Crashworthiness Design Guide," Advanced General Aviation Transport Experiments, AGATE-WP3.4-034043-036 (2002).

The common acronym that relates to crash survivability design factors is CREEP - **c**ontainer (fuselage structure), **r**estraint (seat belts, seats and attachments), **e**nergy management (deformation characteristics), **e**nvironment (objects within occupants' strike zone), and **p**ost-impact factors (fire and egress). The following are some typical crashworthiness design objectives and concerns:

1. Maintaining a protective shell around the occupants by keeping them away from deformed and separated airframe structure during impact, along with providing sufficient energy absorption characteristics. This requires evaluating
   a. Human tolerance to impact along the longitudinal, vertical, and lateral axes
   b. The occupants' strike envelope, which is the potential displacement or flailing of the head, upper torso, arms, and legs (while sitting in an anchored seat and using a proper lap belt and/or shoulder harness)
   c. Sufficient airframe strength to minimize cockpit/cabin volume reduction
      i. Rollover protection preventing a collapsed ceiling
      ii. Minimize inward buckling effects during longitudinal crushing
   d. Delethalization of the fuselage interior
      i. Sufficient attachment strength of components
      ii. Sufficient tie-downs to secure cargo
   e. Minimized earth-scooping effects when the landing gear separates during the crash or during a gear-up landing
      i. Slanted versus sharp-angled edge profiles on lower forward airframe structure causing dragging over versus digging into the terrain, respectively
      ii. Extended structure sticking out of aircraft creating ground plowing (e.g., just the nose gear wheel breaking off and exposing its strut)
   f. Preventing the powerplant from separating from its mounting in front of the cockpit and cabin, which turns into a dangerous obstacle during the crash sequence
   g. Maximized windscreen protection (e.g., from bird or drone strikes)

2. Aircraft seats and restraint systems that minimize *secondary* impacts during deceleration such as head and other flailing injuries. This requires evaluating
   a. Seating orientation – forward, side, or rear facing
   b. Seat strength, anchoring (including seat tracks)
   c. Restraint strength, routing, and anchoring
   d. Energy-absorbing seat structure, cushions, and headrests
   e. Inertia reels locking properly
3. Prevention of post-impact fires – maximize time for the occupants to safely egress before being burned or succumbing to smoke inhalation. This requires evaluating
   a. Fuel tank construction
      i. Smooth shapes with large radii – minimize puncture/tear potential
      ii. Rupture-resistant material
      iii. Foam-fill to reduce atomization during tank rupture (not always practical)
      iv. Anti-spillage vent valves (for rollover accidents)
      v. Flush-mounted drain valves (versus protruding valves that can get torn or ground off during gear-up landings)
   b. Undesirable fuel tank locations
      i. Immediately adjacent to occupiable areas or ignition sources (such as engine or electrical compartments)
      ii. Under heavy masses (such as transmissions and engines)
      iii. In anticipated areas of propeller/rotor blade impact
      iv. Near the bottom of the fuselage where ground contact is likely during gear-up landings
      v. Over landing gears (in case they collapse and puncture the tank during impact)
      vi. In anticipated failure/fractured areas of the wings during a terrain impact (where most of the fuel is typically stored)
   c. Fuel/oil/hydraulic line routing (rigid and flexible)
      i. Similar to undesirable fuel tank locations
      ii. Maximize bend radii
      iii. Minimize exposure to high-tension loads and shearing during impact deformation
      iv. Minimize and seal lines routed through cockpit/cabin area
      v. Avoid turbine rotor burst plane
   d. Routing of electrical wiring to prevent breaks (sparks and arcing) that can ignite a fire or cause an explosion – similar to criteria for fuel/oil/hydraulic line routing
4. Minimize the time needed for occupants to get out of the aircraft during an emergency. This requires evaluating
   a. Location and size of exits (doors and windows)
   b. Ease of opening exits from the inside or outside
   c. Beefing up airframe structure around exits to prevent jamming from deformation during crash landings*

* Good example of this safety issue is addressed in NTSB Report AAR-97/04, DCA97MA009A/B, Runway collision of United Express Flight 5925 and Beechcraft King Air A90, at Quincy Municipal Airport, Quincy, IL, on November 19, 1996.

Crash Load Evaluation

This section deals with *estimating* g-loads sustained by aircraft occupants during a crash, without the benefit of performing specifically designed impact tests. An old saying applies here: "It's not the fall that will hurt you - it's the sudden stoppage." Helpful resources on this subject include NTSB General Aviation Crashworthiness Project, Phases I, II, and III, 1983–1985 and the US Naval Flight Surgeon's Manual, Chapter 24 – Aircraft Accident Survivability.

Three main variables needed to evaluate impact injuries are deceleration distance (D), impact velocity (V), and g-load (G). The collapsing of airframe structure between the terrain/barrier/obstacle and the aircraft occupant absorbs energy that can protect or minimize injuries to the occupants (this typically does not apply to high-speed or inverted impact scenarios). This includes [i] aft, upward, and lateral crushing of fuselages; [ii] chordwise and spanwise crushing, and upward bending of wings; [iii] upward and aft buckling of engine mounts; and [iv] stoking/flexing/collapsing of landing gear. This does not include volumes enclosed by cowlings and fairings unless their structures contain some reinforcement (e.g., longerons and bulkheads within the nose section or engine mounts within engine cowlings or nacelles). We also need to estimate the aircraft impact attitude and breakdown the variables into horizontal (or longitudinal) and vertical components (Figure 5.1).

The following process can be used to approximate crash loads sustained by aircraft occupants and assumes that the occupants were originally seated and using restraints:

1. Obtain aircraft performance and design information:
   a. Operating airspeeds – stall, takeoff, landing, etc.
   b. Normal angle of attack (AOA) during the various operating conditions from flight testing
   c. Weight distribution and center of gravity (e.g., high-wing airplanes in an upright and wings-level attitude during terrain impact have a greater tendency to nose over than a low-wing airplane)

FIGURE 5.1 Components of impact attitude and interrelationships of angles.

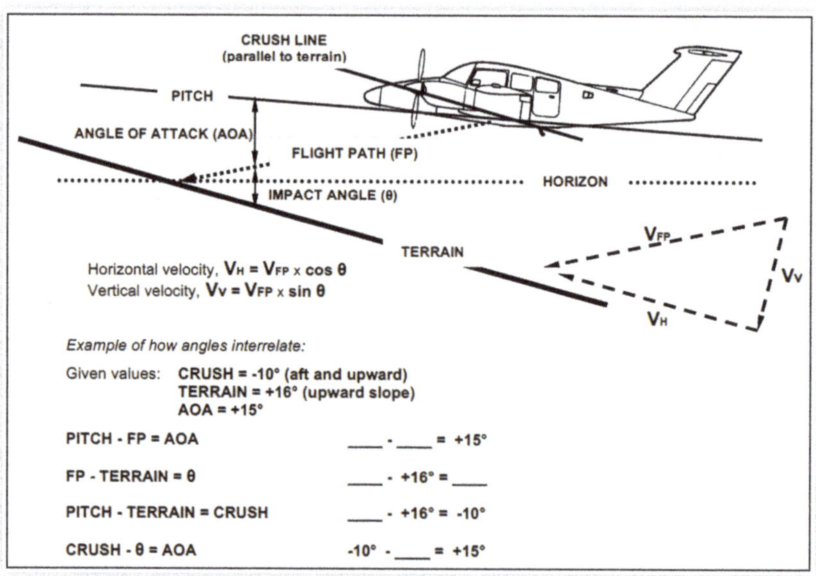

   d. Aircraft dimensions – three-view drawings with fuselage stations, waterlines, and buttock lines
   e. Seating arrangement and locations
   f. Depth of crushable lower fuselage and floor structure
   g. Seat dimensions and configuration
   h. Dynamic test documents for seat certification
   i. Research available crash test data* on similar aircraft configurations (e.g., X° crush line corresponds to a Y° nose-up or nose-down pitch rotation at full crush development or fuselage structure, landing gear, etc. that tend to absorb specific crash loads under various impact attitudes and speeds)
2. Review overall aircraft damage/deformation and derive terrain impact orientation:
   a. Obtain crush line orientation using three-view drawings and wreckage photos – viewed from front, top, and both sides (Figures 5.2 and 5.3)

FIGURE 5.2 Side, top, and front views of Beech C23 Sundowner that show the crush line orientation.†

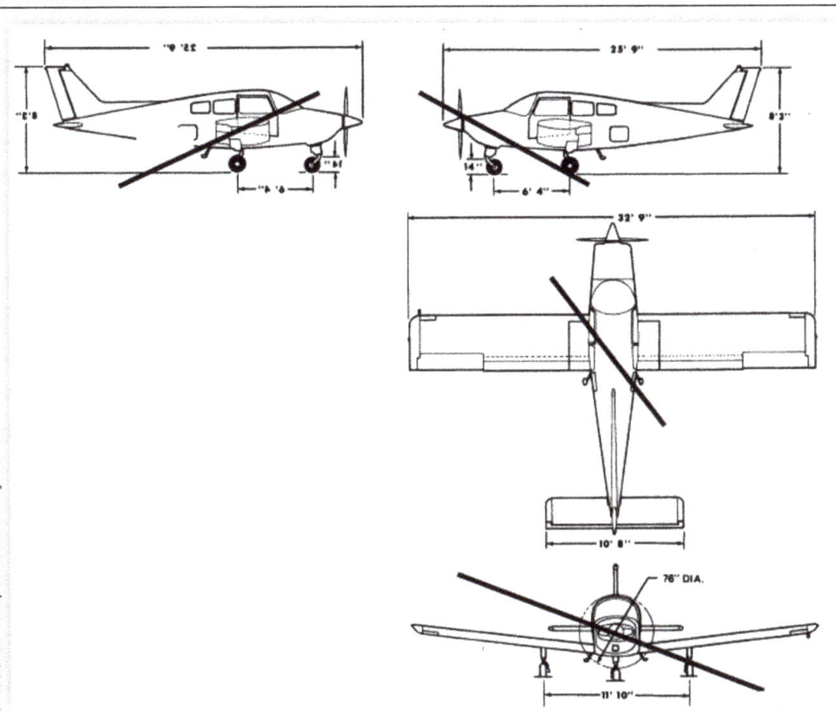

* Actual crash test references include NASA Technical Paper 2083, Correlation and Assessment of Structural Airplane Crash Data With Flight Parameters at Impact, November 1982; A History of Full-Scale Aircraft and Rotorcraft Crash Testing and Simulation at NASA Langley Research Center, January 2006; and Crash Tests of Three Cessna 172 Aircraft at NASA Langley Research Center's Landing and Impact Research Facility, NASA/TM-2015-218987, November 2015.

† Obtained from NTSB General Aviation Crashworthiness Project, as referenced at the beginning of this section.

FIGURE 5.3 Resultant terrain impact orientations derived from crush lines depicted in Figure 5.2.

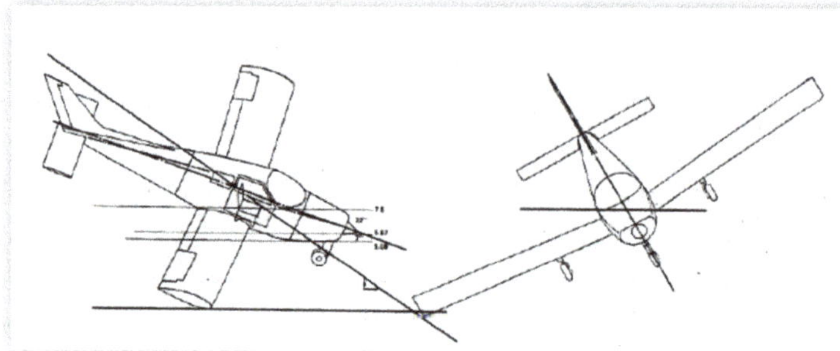

© National Transportation Safety Board

3. Gather witness and first responder statements, emergency rescue reports (from firemen, police, and ambulance), hospital records, and autopsies:
   a. Descriptions of aircraft attitude/orientation during impact[s]
   b. Determine occupant injuries and seat location
   c. Seat restraints being used (or not) – lap belt and type of shoulder harness
   d. Biomechanics analysis of forces required to produce injuries
4. Review terrain information to evaluate initial and final velocities for each stage of the aircraft impact sequence:
   a. Dimensions and locations of ground impact marks (earth scooping, terrain divots and gouges from various parts of the aircraft, propeller slash marks, etc.) – compare with above-noted descriptions of aircraft attitude/orientation during impact
   b. Distance, bearing, and elevations of trees or other obstacles in relation to ground impact marks
   c. Overall condition (dry or wet/muddy soil, hard or spongy ground, rocky terrain, etc.)
   d. Angle and direction of sloped terrain
   e. Estimated friction coefficients for skidding on grass or pavement (without appreciable earth scooping)
5. Evaluate vertical stopping distances (Figure 5.4, Figure 5.5, and Figure 5.6)*:
   a. Depth of ground impact marks

FIGURE 5.4 Components of vertical stopping distance on a small airplane (ground scar not included).†

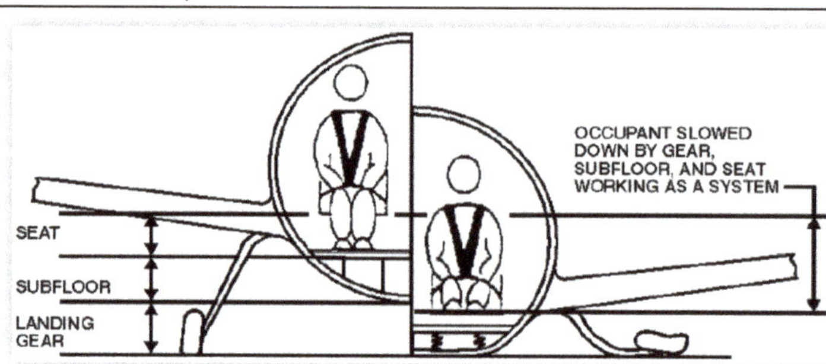

© NASA

* Aircraft impact tests performed by NASA show that the coefficient of restitution during ground impact was equal or close to zero.

† Depiction from "Small Airplane Crashworthiness Design Guide" referenced near the beginning of this chapter.

FIGURE 5.5 Components of vertical stopping distance on a helicopter (ground scar not included).*

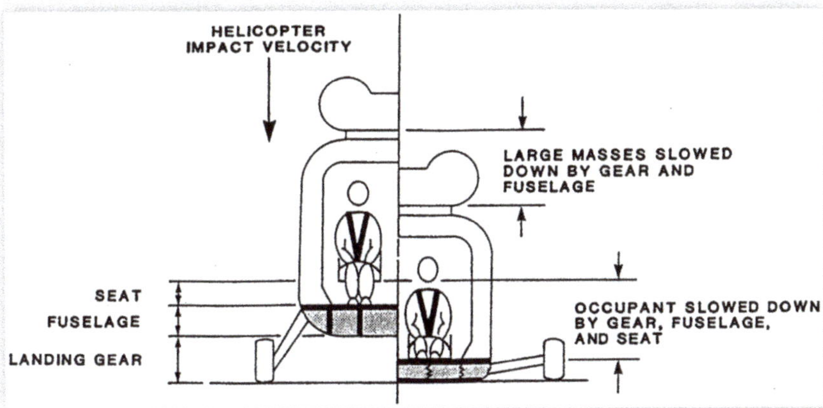

© FAA Technical Center, Report No. DOT/FAA/CT-91/7

   b. Depth of water entry before stopping (low-speed impacts)
   c. Depth/crush distance of subfloor structure
   d. Collapsing distance of landing gear
   e. Collapsing distance of seat structure (including whether seat detached from floor or seat tracks)

6. Evaluate horizontal stopping distances:
   a. Length of ground impact and skid/drag marks
   b. Stopping distance after water entry (low-speed impacts)
   c. Longitudinal crushing distance of airframe structures
   d. Aft displacement of the wings, empennage, and engines
7. Choose a deceleration pulse shape (acceleration versus time) – here are some basic crash conditions:
   a. Rectangular pulse – close to a constant deceleration
      i. Some applications: skidding over terrain without appreciable ground plowing or impacting barriers or ground-arrest systems designed for controlled crash conditions
      ii. Optimal use of crush distance because of maximum velocity change in the shortest distance for a given amount of deceleration, that is, minimal g-load
      iii. Equations: $G = (V_1^2 - V_2^2) / 64.4D$, $t = (V_1 - V_2) / 32.2G$, where G = g-load, t = impulse or duration time, V_1 = entry speed, V_2 = exit speed, and D = deceleration or energy absorption distance

FIGURE 5.6 Cross-section of a Boeing 707 cabin and subfloor, which is indicative of a large transport category airplane – note the much greater available vertical crush distance.

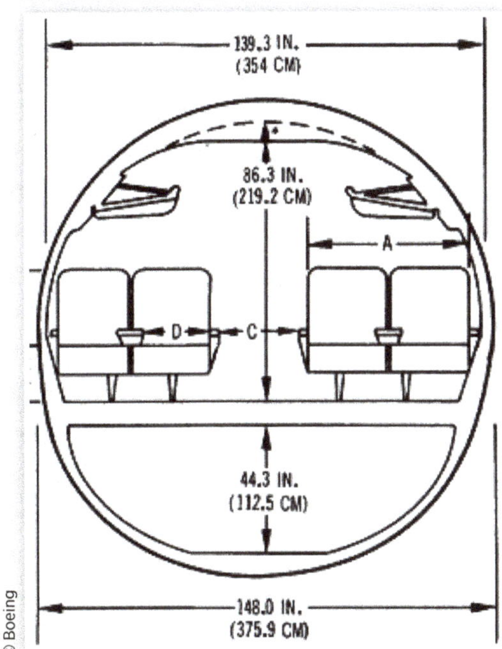

© Boeing

   b. Triangular pulse – continual increase in deceleration
      i. Some applications: impacting and breaking through a stout tree; significant wing structure crushing aft and separating from the airframe; or skidding on terrain with a continual increase in ground plowing
      ii. Equations: $G = (4V_1^2 - 2V_1V_2 - 2V_2^2) / 96.6D$ and $t = 2(V_1 - V_2) / 32.2G$

* Depiction from Rotorcraft Crashworthy Airframe and Fuel System Technology Development Program, FAA Technical Center, Report No. DOT/FAA/CT-91/7, October 1994.

FIGURE 5.7 Human tolerance to longitudinal deceleration (eyeballs forward).

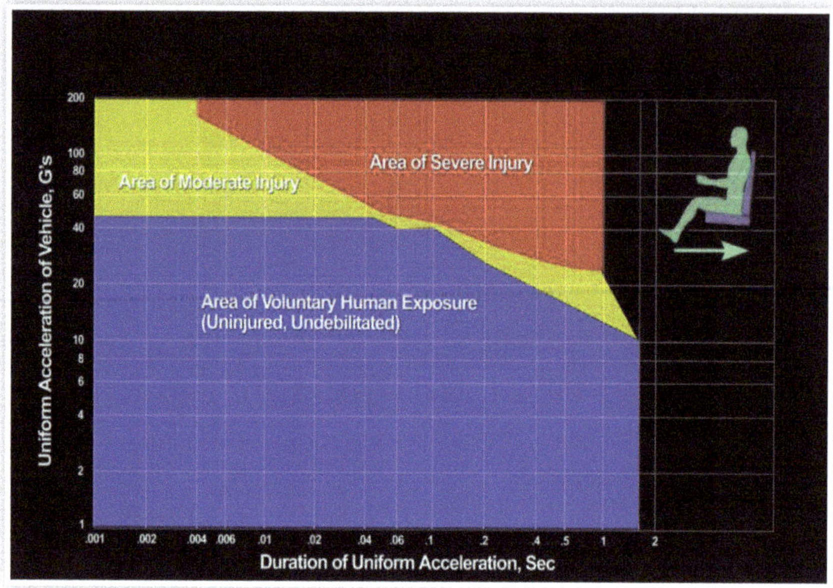

© Don Knutson

c. Symmetrical triangular pulse – increasing and then decreasing deceleration

   i. Application: impacting a barrier or object that does not appreciably give way

   ii. Equations: $G = (V_1^2 - V_2^2) / 32.2D$ and $t = 2(V_1 - V_2) / 32.2G$

8. Refer to longitudinal and vertical g-load charts in Figures 5.7 and 5.8 (derived from crash research data) for evaluating how impact loads and their

FIGURE 5.8 Human tolerance to vertical deceleration (eyeballs down) – yellow shaded area can result in serious back injuries, and NTSB research studies showed a survivability limit of 30 G's, with a velocity change of 50–60 ft/s.

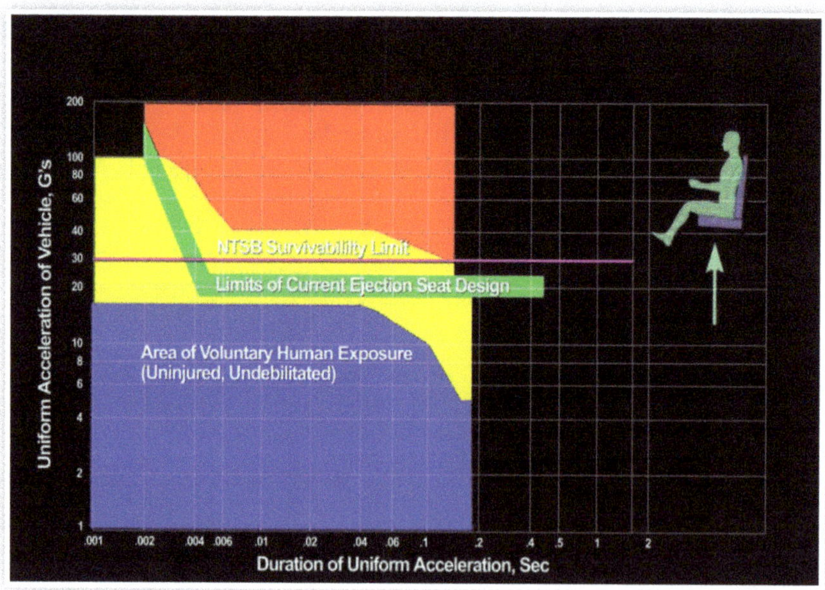

© Don Knutson

> duration can affect aircraft occupants. Remember that age and overall health play a role in injury tolerance, so the upper limits of "uninjured, undebilitated" could result in serious injuries, and the serious injuries can turn into fatal injuries. Also, how well someone remained restrained in their seats and the severity of aircraft structural damage significantly affects the survivability of a crash event. For example, Figure 5.7 shows that serious or fatal injuries can occur at a longitudinal deceleration of 45–50 G's, with a 0.08-s duration at optimal occupant restraint conditions; however, NTSB research studies (referenced at the beginning of this section) showed a survivability limit of 35 G's, with a velocity change of 60–70 ft/s when taking into account severe secondary impacts such as with the cockpit instrument panel and/or control yoke (wheel) assembly.

Occupants located in forward or aft-facing seats along the sides of the cockpit or cabin can most likely survive a *lateral* impact of at least 20 G's for 0.1 s during a crash*. However, these circumstances can result in serious head injuries due to sideways flailing and no airbag deployment or structure designed to attenuate the head impact loads. Research has shown that neck injuries from sideways whiplash (as shown in Figure 5.9) are more likely to occur than head impact injuries to restrained occupants in side-facing seats during an aircraft crash. Also, serious neck injuries to occupants wearing restraints in side-facing seats can occur even when the aircraft complies with emergency landing dynamic conditions noted in FAR 25.562 (a minimum longitudinal velocity change of 44 ft/s, with a 10° yaw either right or left of the longitudinal axis, and a minimum peak of 16 G's reached in not more than 0.09 s).†

In summary of the aforementioned, the majority of g-loads sustained by aircraft occupants during a crash will involve a symmetrical triangular pulse shape. However, airframe material configurations that efficiently absorb energy during crashes, such as sandwiched composites‡ (e.g., honeycomb and foam-core airframe structures), can result in overall crash profiles somewhere between a rectangular (ideal) and symmetrical triangular pulse shape (Figure 5.10). Also, even though composite aircraft structures tend to be more rigid than aluminum airframes, they still can incorporate innovative design concepts that provide

FIGURE 5.9 Sequence of trunk and head responses (whiplash) while seated during a lateral impact and no head contact against an object.§

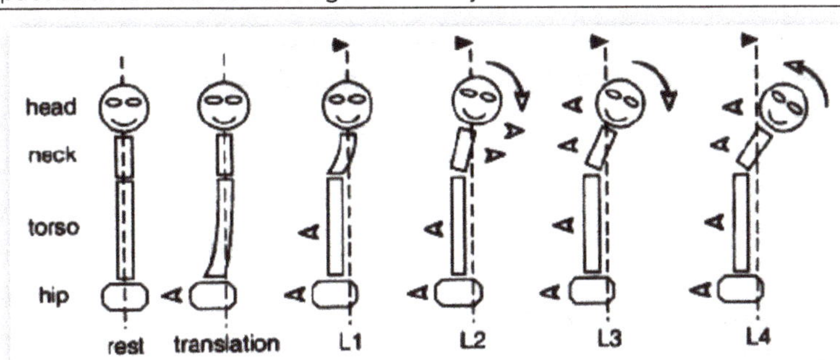

* "Aircraft Crash Survival Design Guide" referenced near the beginning of this chapter.

† FAA CAMI Report DOT/FAA/AM-07/13, Assessment of Injury Potential in Aircraft Side-Facing Seats Using the ES-2 Anthropomorphic Test Dummy (2007).

‡ FAA Aviation Maintenance Technician Handbook-Airframe (FAA-H-8083-31A), Volume 1, Chapter 7, regarding sandwich composite construction and materials.

§ "Variability in the control of head movements in seated humans: a link with whiplash injuries?" Journal of Physiology 532.3, pp. 851-868, Vibert, N., MacDougall, H.G., Waele, C. De, Gilchrist, D.P.D. (2001).

FIGURE 5.10 Rectangular and symmetrical triangular pulse shapes.

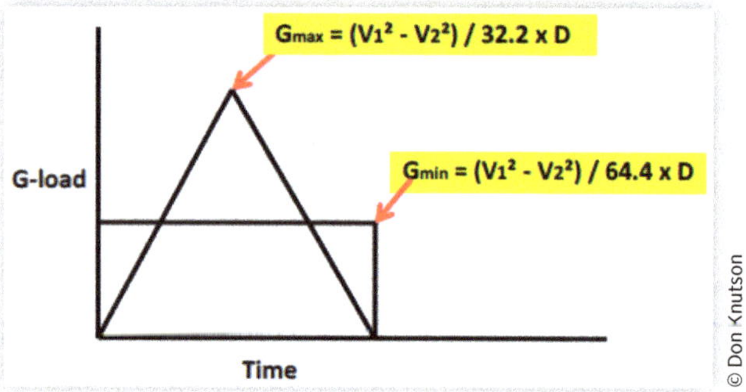

crumple zones with notably improved energy absorption characteristics* (e.g., notched corners, corrugated webbing, and less stiff laminate layups in strategic locations that provide a controlled collapsing of structure during impact).

Let's look at an example of crashworthiness analysis – assume the following circumstances†:

- Beech 99 Airliner on an ILS approach in fog and relatively calm wind.
- Copilot stated that the "bottom dropped out" while on final approach.
- Airplane crashed on a level ice-covered bay and slid onto the shore with no forward obstacle impacts.
- Pilot in left front seat sustained serious injuries, including compression fractures of the vertebrae and broken legs.
- Copilot in right front seat sustained minor injuries.
- Airframe crush lines measured directly from a three-view drawing showed a roll angle of 21° left wing down and a 9° nose-down pitch attitude (Figures 5.11 and 5.12)‡.
- Wing flaps were set at 30% with landing gear down/extended.
- Propeller blade damage was consistent with low engine power settings.

Terrain impact speed and angle can be estimated by considering the following:

- Copilot statement that "bottom dropped out" is consistent with the airplane near stall speed at initial impact.
- Stall speed = 86 knots true airspeed (KTAS), with power off, gear down, and 30% flaps.
- During the noted stall conditions, the AOA = -12° and pitch = -9° (nose down).

FIGURE 5.11 Side, top, and front views of Beech 99 Airliner, which shows relative crush line orientations.

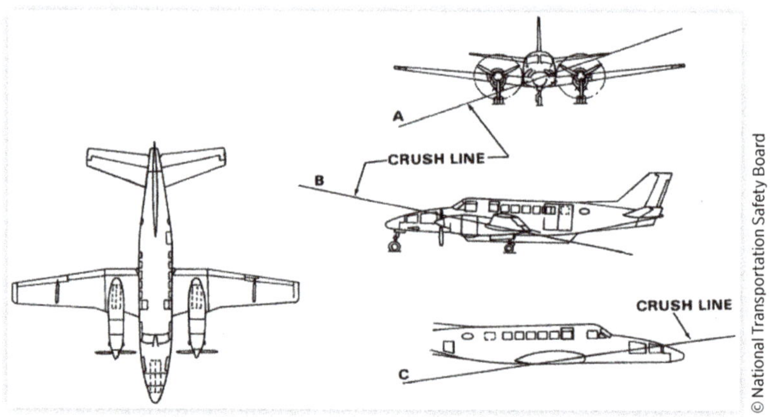

* Cronkhite, J. D. and Berry, V. L., "Crashworthy Airframe Design Concepts - Fabrication and Testing," NASA CR3603, Washington, D.C., September 1982.

† Derived from NTSB Report NYC82AA028, windshear encountered during landing approach, at Groton, CT, on February 1, 1982, N451C, Beech 99 Airliner.

‡ Obtained from NTSB General Aviation Crashworthiness Project, as referenced at the beginning of this section.

FIGURE 5.12 Resultant terrain impact orientations derived from crush lines depicted in Figure 5.11.

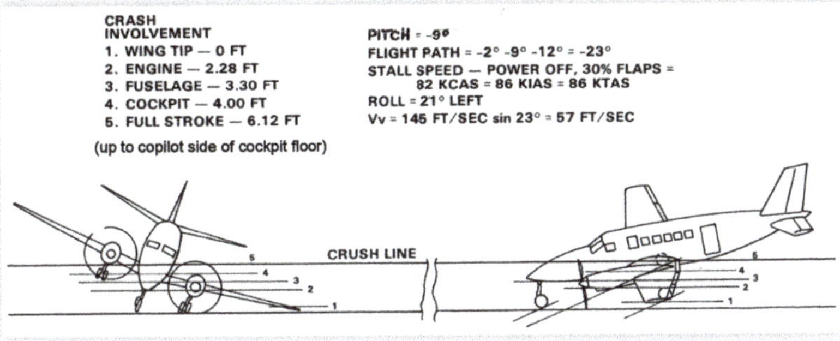

- Evaluation of available NASA crash test data on similar low-wing airplane configurations indicates that the resultant crush line orientations would correspond to a 2° nose-up pitch rotation at full vertical crush development.
- Flight path angle = AOA + nose-down pitch - pitch rotation after impact = -12° - 9° - 2° = -23°.
- Impact angle = flight path angle - terrain angle = -23° - 0° = -23°.
- Flight path velocity, V_{FP} = 86 KTAS = 145 ft/s.
- Vertical velocity, $V_V = 145 \times \sin 23° = 57$ ft/s.

Forces on occupants can be estimated by considering the following:

- Based on the airplane's orientation while impacting terrain (a relatively rigid barrier), we can focus on the vertical g-loads that caused the injuries.
- Visualization of crush lines shows that primary structure (landing gear) will start absorbing energy upon initial impact and all of the crushing distance will be utilized (6.12 ft).
- Stroking distance to left side of cockpit, D_1 = 4.00 ft, with chosen symmetrical triangular pulse shape equations, $G = V^2/32.2D$ and $t = 2V/32.2G$.
- Vertical g-load, $G_{v1} = 57^2/(32.2 \times 4.00) = 25.2$*, and t = 0.14 seconds.
- Stroking distance to right side of cockpit, D_2 = 6.12 ft.
- $G_{v2} = 57^2/(32.2 \times 6.12) = 16.5$ and t = 0.21 s.
- Estimated g-loads are consistent with injuries to pilot vs. copilot – the pilot was seated within the crush zone and serious back injuries have been found to usually occur with a 20+ vertical g-load.

Using G-Load Survivability Limits to Look Deeper

Understanding the basic methodology of estimating g-loads and the empirical limitations to survivable impacts helps us to further evaluate possible circumstances leading to an accident. For example, a twin-engine airplane crashed into a remote mountainside at an ~8,000-ft elevation, and the airplane's intended destination was an airport on the other side of the mountain.

* The vertical g-load calculations differ from the referenced NTSB research document because I believe the lateral collapsing of the nose and left main landing gears along with the upward deflection of the left wing structure helped attenuate the impact loads during initial terrain contact more than the NTSB anticipated.

Weather studies indicated that the airplane had an ~15-knot headwind during initial terrain contact and was not affected by mountain wave turbulence (e.g., downdrafts). Due to broken clouds along the mountains being reported at the 8,000-ft elevation, the government agency in charge of the investigation concluded that the cause of the accident was CFIT* while the airplane was attempting to fly over the mountains. No radar data or known radio communications were available. Three people were found on board the airplane and seriously injured by blunt force trauma (from frontal impact deceleration loads). The three occupants were reportedly found to be responsive by local villagers who first arrived on-scene; however, only the pilot in the right front seat ultimately survived. The accident scene documentation and wreckage inspection revealed that the airplane crash sequence involved four main stages (refer to Figure 5.13):

- From points 1 to 2 – initial terrain contact (Figure 5.14) involving the bottom of the right engine nacelle skidding ~20 ft on a grassy hillside (which sloped laterally upward in relation to the right side of the airplane).
- From points 2 to 3 – immediately descended into a valley ~30 ft and impacted a tree ~300 ft (horizontal distance) from the initial terrain contact.
- From points 3 to 4 – 4-ft-wide portion of the right wing separated just outboard of the engine nacelle while impacting a tree.
- From points 4 to 5 – after the right wing separation, the fuselage nose section impacted the ground in a steep nose-down attitude and crushed aft to the cockpit area.

The overall analysis was started from point 5 and worked upstream to the initial terrain contact area because the three airplane occupants were initially found alive and the pilot in the right front seat ultimately survived the crash – this gives us a point of reference regarding maximum g-loads imparted on the airplane occupants. In this case, NTSB research studies showed a survivability limit of 35 G's along the aircraft's longitudinal axis when considering secondary impacts (and both front seat occupants impacted their control yoke as well as the instrument panel).

FIGURE 5.13 Diagram with points of interest and governing equations based on their impact pulse shapes (g-load versus time) used to evaluate crash dynamics at accident site.

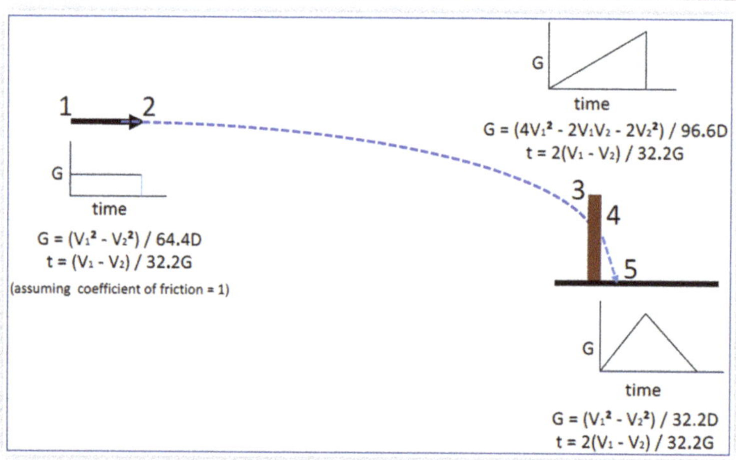

© Don Knutson

* Per FAA Advisory Circular 61-134, CFIT Awareness, CFIT occurs when an airworthy aircraft is flown, under the control of a qualified pilot, into terrain (water or obstacles) with inadequate awareness on the part of the pilot of the impending collision. Per FAA Instrument Procedures Handbook (FAA-H-8083-16), CFIT is an event in which a normally functioning aircraft is inadvertently flown into the ground, water, or an obstacle and is caused by the flight crews' lack of vertical and/or horizontal position awareness.

FIGURE 5.14 Initial airplane impact area consistent with the lower portion of the right engine nacelle skidding along the grassy terrain.

From Points 4 to 5

The wreckage examination revealed that the fuselage nose section exhibited primarily longitudinal/aft crushing (Figure 5.15), which provided an energy-absorbing crush distance of ~5 ft. Assuming 35 G's would not be exceeded and the crash dynamics had a symmetrical triangular pulse shape results in an estimated impact velocity of ~75 ft/s or ~44 knots (using the governing equation noted in Figure 5.13).

FIGURE 5.15 Fuselage nose section was crushed aft and upward during the steep nose-down impact (point 5 in Figure 5.13).

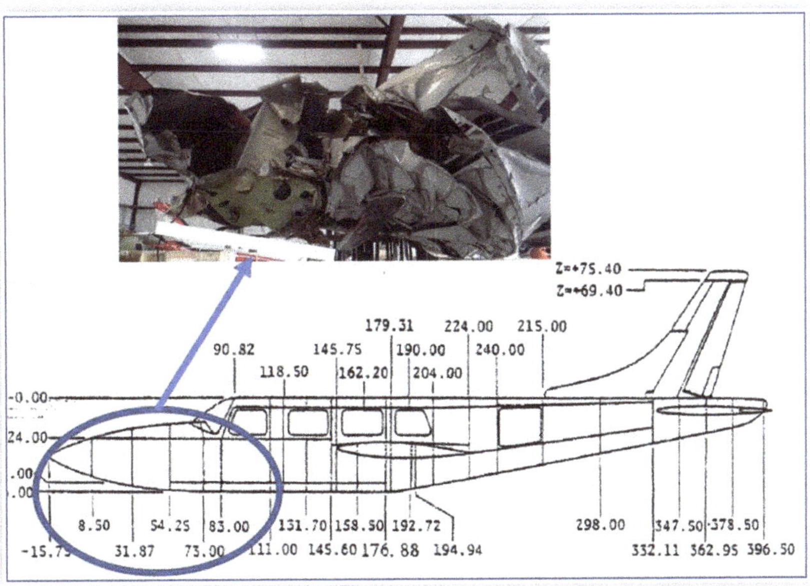

From Points 3 to 4

Considering that a 4-ft-wide section of the right wing structure was torn off during the tree impact, the crash acceleration versus time was approximately a triangular pulse shape (continuing increase in deceleration). Again, assuming that 35 G's would not be exceeded during the tree impact and using 75 ft/s as the exit velocity, the estimated entry impact velocity would have been ~100 ft/s or ~59 knots.

From Points 2 to 3

After the airplane skidded off the hillside and immediately descended, its vertical speed component would have increased due to gravity ($V_v^2 = 2g_c d_v$, where $d_v = 30$ ft and $g_c = 32.2$ ft/s^2). Since the vertical speed would have increased ~44 ft/s and the tree impact speed along the flight path could not exceed ~100 ft/s to remain (assumed to be) survivable, it stands to reason that the airplane's exit velocity coming off the skid mark was at a stall speed. Therefore, a conservative groundspeed of 100 ft/s was assumed to be the exit velocity coming off the skid mark (based on stall speeds at a weight between 4,500 and 5,000 pounds noted on Figure 5.16, and the reported ~15-knot headwind).

From Points 1 to 2

Skidding ~20 ft along the hillside resulted in an approximate rectangular pulse shape. NTSB performed crash tests using twin-engine airplanes weighing close to the subject accident airplane and impacting soil with their landing gear retracted (similar to the accident airplane), which resulted in ground sliding friction coefficients (μ) of 0.42. In this case, a conservative μ of 1 is used due to a minimal amount of ground plowing that occurred, which results in the equation $V_1^2 = 1{,}288G + 10{,}000$, where V_1 is the entry speed and G is the average deceleration of the airplane during the skid.

The skid mark and wreckage damage (fuselage nose section and right wing leading edge did not ground plow into the soil) was consistent with the airplane maintaining an approximate nose-level attitude during the initial terrain impact. Also, an exit groundspeed of 100 ft/s along ~20 ft results in the airplane contacting terrain for no more than a brief 0.2 s, which should not have generated any substantial deceleration loads when considering the above-noted impact orientation. Assuming a very conservative 5 G's with the equation noted in the previous paragraph results in $V_1 = 128$ ft/s or 76 knots. This is inconsistent with a CFIT scenario because this speed is obviously closer to a setup for an emergency landing (near the stall speeds noted on Figure 5.16).

Another good example of considering the survivable impact limits to evaluate the sequence of a crash is from an accident presented in Chapter 4 involving an airplane that lost engine power in-flight, and was forced to land in a field and the engine assembly and nose landing gear wheel separated during the impact sequence. The pilot and passenger in the front seats survived the crash but sustained serious back injuries. These findings *possibly* suggested that the pilot could have flared the airplane at too steep of a nose-up attitude during the emergency landing, which resulted in a stall and abrupt nose-down impact with terrain. NTSB crashworthiness research shows that the survivable upper limit of *vertical* acceleration is 30 G's; therefore, we can further determine whether the stall theory has merit by calculating the vertical loading on the airplane occupants via establishing an effective crush distance and impact velocity (as outlined earlier in the chapter). If the calculated vertical g-loads show more than 30 G's, then we need to consider

FIGURE 5.16 Stall speed chart from subject accident airplane's Pilot's Operating Handbook.

CONDITION:

1. ZERO THRUST

WEIGHT POUNDS	CONFIGURATION		BANK ANGLE							
			0°		20°		40°		60°	
			KNOTS		KNOTS		KNOTS		KNOTS	
	FLAPS	GEAR	IAS	CAS	IAS	CAS	IAS	CAS	IAS	CAS
5500	UP	UP	83	76	85	78	90	87	108	107
	20°	DOWN	77	70	78	72	83	80	99	99
	45°	DOWN	74	67	75	70	78	76	95	96
5000	UP	UP	79	72	82	75	86	83	103	102
	20°	DOWN	74	67	75	69	79	76	95	95
	45°	DOWN	70	63	71	66	75	73	89	90
4500	UP	UP	76	69	77	70	81	78	97	96
	20°	DOWN	70	63	71	65	75	72	90	90
	45°	DOWN	68	61	68	63	72	70	85	86
4000	UP	UP	71	64	74	67	77	74	92	91
	20°	DOWN	67	60	68	62	72	69	84	84
	45°	DOWN	64	57	64	59	67	65	80	81

NOTE

1. ZERO INSTRUMENT ERROR ASSUMED.

other possibilities of how the aft/upward crushing on the lower forward portion of the fuselage cabin area occurred (e.g., the fuselage could have collided with a terrain obstacle or the separated engine downstream of the initial ground impact at a slower speed).

Emergency Egress Issues

Getting out of an aircraft after it crashed can turn into a daunting task. Some of the more typical problems people have with egressing from an aircraft are

- Fires from ignited/leaking fuel (e.g., ruptured tanks atomizing fuel on a heat source)
- Water ditching when the aircraft abruptly flips upside down and/or rapidly submerges causing disorientation and panic
- Terrain impact injuries to the head, neck, back, legs, or feet that render the occupants unconscious or immobile
- Airframe deformation during the crash that prevents occupants from readily opening the cabin doors or exits (refer to the first footnote of this chapter)
- Cabin smoke from onboard fires preventing the occupants from reaching or finding the exit from smoke inhalation and their eyes burning (on larger fuselage cabin and transport aircraft)

Water ditching procedures are one of the areas in GA that a significant number of pilots are not readily prepared nor trained to encounter. Many water ditching accidents do not result in significant aircraft damage or impact-related injuries. However, the pilots and/or passengers become confused, shocked, and panicky when almost immediately being inverted and/or submerged during the water impact (especially in the dark and when the water is cold*). This significantly delays their ability to locate a door handle and open the door to escape. In many cases, these occupants have drowned simply because they were not prepared for the emergency, which could have been avoided via a preflight safety briefing by the pilot-in-command (PIC) and the PIC having been properly trained on ditching procedures. Proper ditching egress procedures on small aircraft include

- Unlatching the doors before touchdown
- Releasing the seat restraints *after* the cabin comes to rest in the water
- Maintaining awareness of the cabin exit locations *before* possibly encountering low visibility conditions
- Maintaining access/reach to a life preserver, immersion suit, or raft (before the aircraft sinks)
- Inflating the flotation devices when clear of the aircraft
- Getting everyone on a raft and out of the water as soon as possible to avoid hypothermia

With this in mind, wreckage inspections involving water ditching accidents should include the following:

- Overall crushing damage to estimate the crash loads sustained by the occupants (using the procedures noted earlier in this chapter) – survivable, serious, or minor?
- Obtain reports on occupant impact injuries – pelvic and upper torso bruising from wearing a lap belt and shoulder harness, head or neck injuries from impacting the control yoke or instrument panel, or back injuries from excessive vertical impact loads (e.g., aircraft stalling too high over the water)?
- Aircraft equipped with life preservers, immersion suits, or raft and used by the occupants? If installed, were the occupants capable of accessing them in a reasonable amount of time?†
- Emergency briefing card[s] found in aircraft? If so, were they reasonably available to the passengers?
- Cockpit and cabin doors being unlatched before or after water impact?
- Cockpit or cabin exits being deformed during water impact that could have resulted in door latch damage and being difficult to open?
- Airframe structure wedged against the cockpit or cabin exits making them difficult to open?

---

* "Survival in Cold Waters: Staying Alive," Transport Canada Report TP13822E, January of 2003, which provides information for fishermen and emphasizes the value of immersion suits.

† Keep in mind that government regulatory agencies such as the FAA may not require aircraft operators, carrying passengers for hire and based at airports located on a waterfront, to equip their aircraft with flotation aids nor require their pilots to give preflight safety briefings for water ditching procedures to the passengers. One of the reasons is because the regulators can deem airplanes to be in gliding distance to land if engine power is lost, regardless of the fact that single-engine airplanes are regularly operating well under a typical traffic pattern altitude (1,000 feet) over water during their initial climb-out and while heading away from the airport. In other words, what happens when the airplane loses engine power shortly after takeoff and is over the water with insufficient altitude for the pilot to turn back toward the airport (faced with the "impossible turn")?

CHAPTER 6

In-Flight Breakups

An in-flight breakup is when a part or parts of an aircraft separate during flight without the direct result of impacting another aircraft or obstacle. Causes and contributing factors that lead to in-flight breakups include:

- Performing aerobatic or high-speed maneuvers with abrupt/excessive control movements
- Loss of visual/attitude reference (e.g., spatial disorientation and attitude indicator malfunctions*)
- Flight into severe weather (e.g., severe turbulence, thunderstorms, and icing)
- Control surface flutter (e.g., unbalanced flight control surfaces and free trim tabs)
- Fatigued airframe structure (e.g., poor maintenance and aging aircraft)
- Pilot incapacitation (e.g., drugs, hypoxia and carbon monoxide poisoning)
- In-flight fire or smoke in cockpit
- Explosions (e.g., fuel tanks and bombs)

These types of wreckage reconstructions can be complex and require the help of experienced investigators who have been properly mentored and understand the fundamentals of analytical dynamics, aerodynamics, and aeroelasticity.

* DOT/FAA/AM-02/19, General Aviation Pilot Performance Following Unannounced In-Flight Loss of Vacuum System and Associated Instruments in Simulated Instrument Meteorological Conditions, provides evidence of how critical and dangerous it is when a pilot loses attitude reference.

Behavior of Separated and Falling Parts

Initial information to obtain during an in-flight breakup investigation includes any available accident pilot communications with ATC or other pilots/aircraft flying in the proximate airspace, ADS-B/GPS/radar (latitude, longitude, altitude, direction, and airspeed) data, meteorological conditions, and the applicable Pilot's Operating Handbook and Airplane Flight Manual Supplements (e.g., for limitations, emergency procedures, performance data, and safety warnings). We want to understand whether loss of control or structural failure occurred first, the parts failure/separation sequence, and what altitude/direction/attitude/speed the aircraft was flying at the time of the breakup. Trajectory* analysis can help with these endeavors even when no ADS-B, radar, or GPS data is available. Accurately documenting the locations (latitude, longitude and terrain elevation), weights, and dimensions of each part of the wreckage during the on-site investigation phase is crucial.

Conditions of ballistic parts separating from an aircraft in flight (Figure 6.1) that remain constant are the gravitational acceleration (32.2 ft/s^2 or 9.81 m/s^2) along with their weight and dimensions (assuming they do not further come apart or deform during the free fall). Conditions subject to change are the altitude, direction, attitude, airspeed (until reaching terminal velocity), and the wind speed and direction. The object's frontal area and drag coefficient also changes while tumbling through the air, but can be assumed to remain constant during a trajectory analysis by evaluating the object's weight distribution (heavy end oriented downward) and likely axis of rotation, and then estimating the effective frontal area and average drag coefficient (Figure 6.2). If we anticipate that the falling object will tumble about an axis, then we can simply add the presented frontal planform areas together ($S_1 + S_2$) and then multiply by a factor of 0.637 to estimate a *mean* frontal area†.

FIGURE 6.1 Basic dynamic conditions on a piece of wreckage involved with an in-flight breakup.

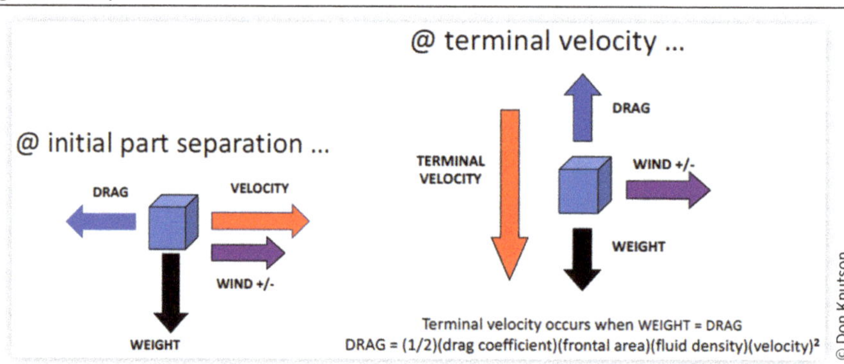

* A trajectory is the path that a moving object follows through space as a function of time. Analysis of the object's flight path is used to locate pieces that separated from an aircraft in-flight with or without resulting in a crash (e.g., separation of the wings and tail section as well as just searching for a lost engine cowl or parts from an uncontained turbine engine failure).

† "Trajectory Analysis in Aircraft Accident Investigation," Report LR-310 (1961), National Research Laboratories, Ottawa, Canada, notes that the mean frontal area, $S_e = 0.637\ (S_1 + S_2)$, was derived from the assumption that a falling and rotating (or tumbling) body will present changing areas to the stream flow that repeats at each quarter revolution, which results in a total area of $[S_1 \sin\xi + S_2 \cos\xi]$, and then approximated to a mean effective area during rotation, $S_e = 2\int_0^{\pi/2}[S_1 \sin\xi + S_2 \cos\xi]\,d\xi/\pi$. NTSB investigations of in-flight breakups have used factors of 0.632 and 0.635 for separated parts subject to rotation.

When practical, drop tests with the subject or similar pieces of wreckage at specified altitudes and atmospheric conditions are the preferred way to evaluate and verify the parts' drag coefficient and motion through the air (especially for irregularly shaped objects). In-flight breakups that resulted from encountering severe weather such as thunderstorms with powerful updrafts and downdrafts can sometimes make it difficult to replicate the ballistic behavior of the thrown/falling parts.

Let's look at an example of calculating the effective frontal area (perpendicular to the direction of travel) of an aircraft part. A fuselage and powerplant assembly, with most of

FIGURE 6.2 Airflow over common shapes with corresponding drag coefficients and Reynolds numbers (obtained from Report LR-310 (1961), National Research Laboratories, Ottawa, Canada).

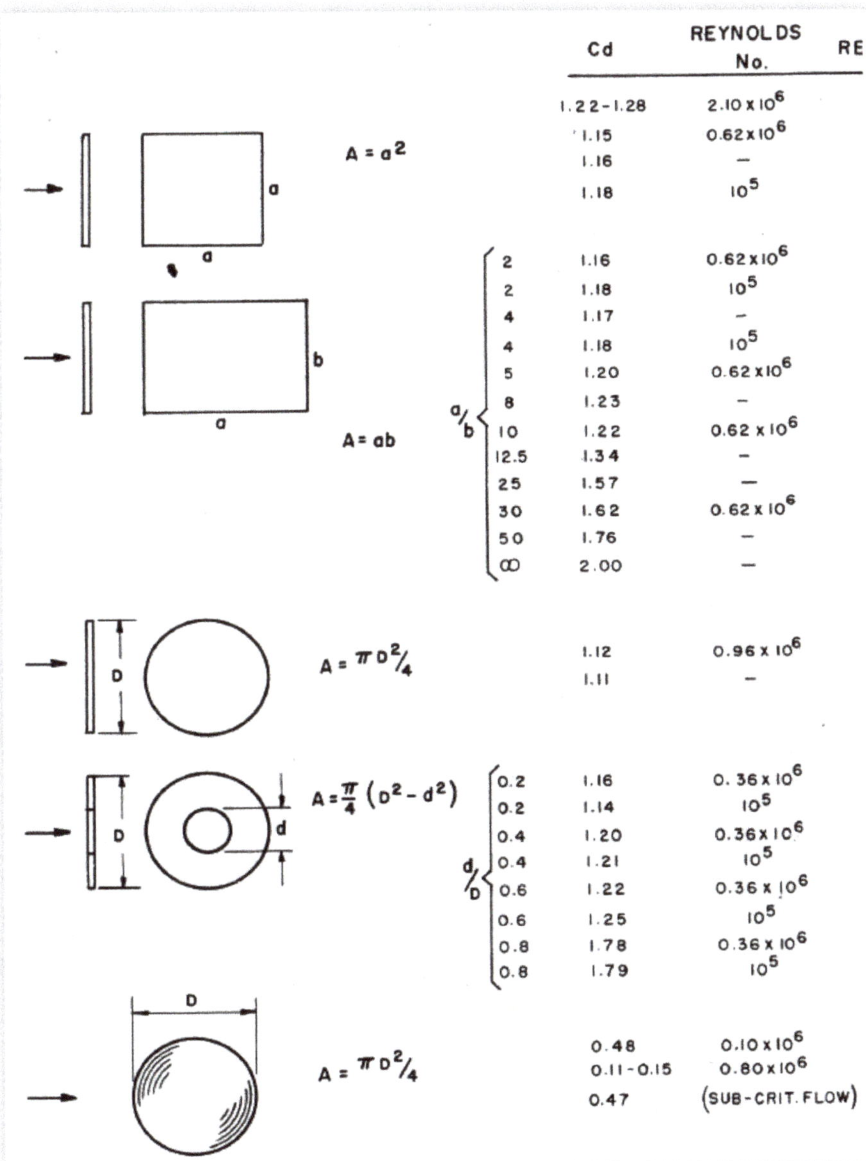

FIGURE 6.2 *(Continued)* Airflow over common shapes with corresponding drag coefficients and Reynolds numbers (obtained from Report LR-310 (1961), National Research Laboratories, Ottawa, Canada).

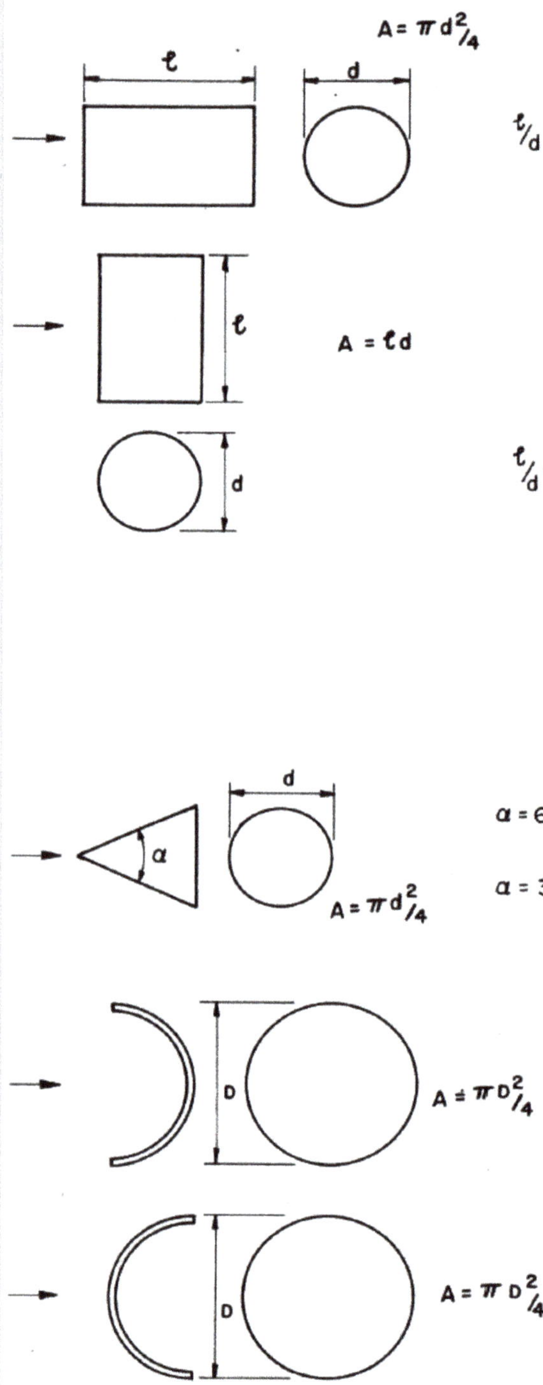

Shape	ℓ/d or α	Cd	REYNOLDS No.
Cylinder, axis parallel to flow ($A = \pi d^2/4$)	0.5	1.00	0.36×10^6
	1.0	0.84	0.36×10^6
	2.0	0.76	0.36×10^6
	4.0	0.78	0.36×10^6
	6.0	0.80	0.36×10^6
	7.0	0.88	0.36×10^6
Cylinder, axis normal to flow ($A = \ell d$)	1	0.64	0.09×10^6
	1	0.91	—
	1	0.71	10^5
	2	0.68	0.09×10^6
	2	0.85	—
	2	0.71	10^5
	4	0.87	—
	4	0.79	10^5
	5	0.76	0.09×10^6
	7	0.99	—
	7	0.82	10^5
	10	0.80	0.09×10^6
	10	0.88	10^5
	20	0.92	0.09×10^6
	40	0.98	0.09×10^6
Cone ($A = \pi d^2/4$)	$\alpha = 60°$	0.51	0.27×10^6
		0.51	—
		0.50	—
	$\alpha = 30°$	0.33	0.27×10^6
		0.34	—
Hemispherical shell, convex side to flow ($A = \pi D^2/4$)		1.44	0.40×10^6
		1.42	0.50×10^6
		1.35	—
		1.42	—
Hemispherical shell, concave side to flow ($A = \pi D^2/4$)		0.36	0.40×10^6
		0.34	0.50×10^6
		0.41	—
		0.38	—

FIGURE 6.2 (*Continued*) Airflow over common shapes with corresponding drag coefficients and Reynolds numbers (obtained from Report LR-310 (1961), National Research Laboratories, Ottawa, Canada).

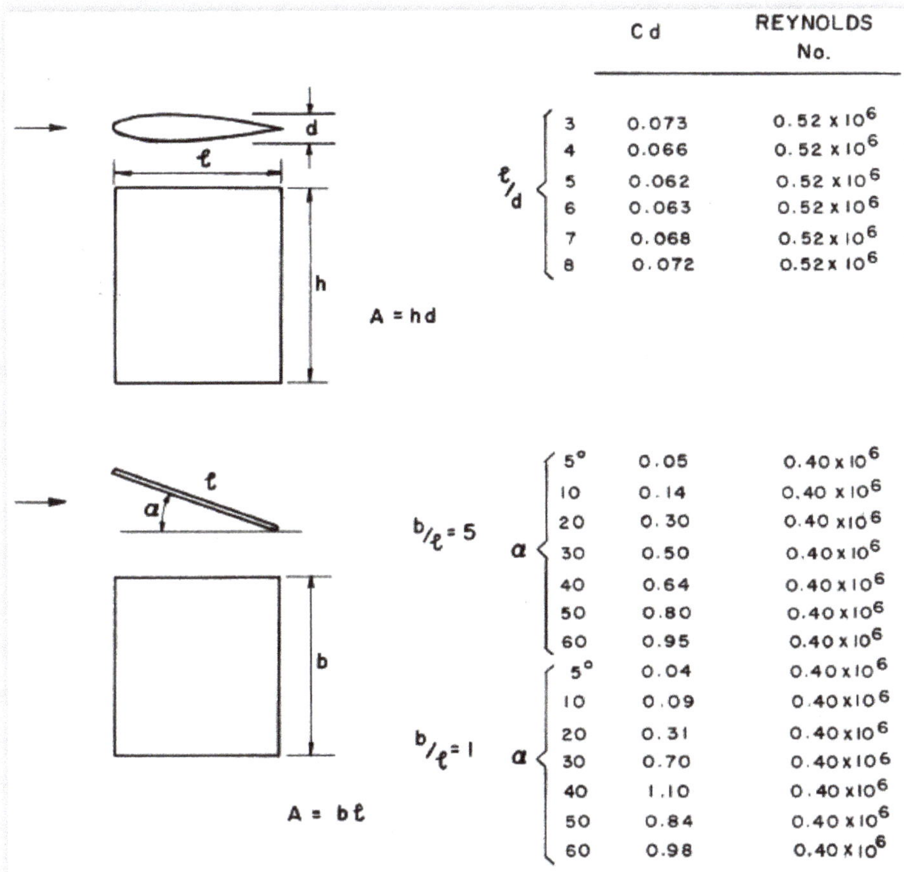

both wings and the aft fuselage and tail section, separated during an in-flight breakup (Figure 6.3). The upper surface area of the fuselage and engine was ~20 ft long × 5 ft wide, and the remaining wing area (excluding fuselage area between both wings) was ~13 ft spanwise × 4 ft chordwise. The weight distribution and resultant airframe crush damage during terrain impact indicates the general orientation that the part fell, which gives us an initial estimated frontal area. In this case, the heaviest part of the airplane was toward the front that favored a nose-first impact, and the overall part fell while in an inverted nose-down attitude somewhere between 30° and 45° (assumed to have no appreciable tumbling). This gives us an initial estimated effective frontal area between 76 and 107 ft^2 (total square feet × sin [inverted nose-down angle]).

Another example is a separated wing section from the aforementioned fuselage (Figure 6.4), with dimensions of ~15 ft long × 3 ft wide × 0.5 ft thick (average along its span). The wing was suspected to have fallen like a maple leaf with its inboard section being the heaviest portion (oriented downward) and rotating about its spanwise axis. This gives us an initial estimated mean frontal area of 0.637([15 × 3] + [3 × 0.5]) = 29.6 ft^2.

FIGURE 6.3 Views of an inverted fuselage wreckage with its powerplant and without its aft fuselage and tail section.*

FIGURE 6.4 Left wing section that separated from fuselage depicted in Figure 6.3.

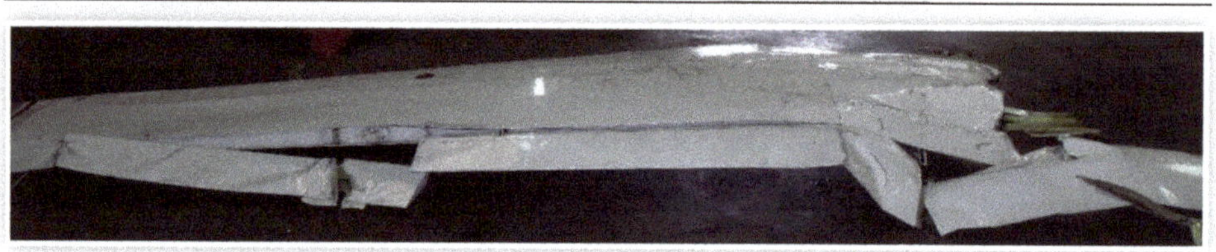

Calculating the ballistic coefficients (B) of each piece of wreckage helps estimate how each piece will be affected by the winds. $B = W/(C_d \times S)$, where W = weight, C_d = drag coefficient, and S = effective or mean frontal area (depending on whether the part tumbles). Good examples of pieces with low ballistic coefficients are styrofoam balls and sheets of paper - both are highly affected by wind drift. Conversely, examples of high ballistic coefficients are bowling balls and aircraft engines – they have a concentrated mass and shape that is not as easily blown around (Figure 6.5).

* Obtained from NTSB Accident Docket ERA14FA058, Piper PA-46-310P, N87NF, in-flight breakup near Dawsonville, GA, on December 2, 2013.

A general process for trajectory analysis includes (1) establishing (or assuming) the aircraft's location within a horizontal plane, altitude (vertical component), heading (horizontal component), and horizontal and vertical velocities (which provides the launch or flight path angle) at the time of the in-flight breakup; (2) breaking down the time intervals between calculations into small segments due to the integrating nature of the calculations; (3) inputting the wind speed and direction at altitude during each time segment; (4) inputting the separated object's weight, anticipated frontal area, and drag coefficient (that can also be obtained from drop test data) along with the air density at altitude to calculate the pressure drag during each time segment; (5) calculating the separated object's acceleration, velocity, and location in horizontal and vertical components during each time segment* (terminal velocity is reached when the object's weight equals its drag force); (6) calculating wind drift during each time segment; and (7) going back to step 3 until the altitude equals the terrain elevation. Due to variable winds aloft and aerodynamic behavior of the thrown/falling objects while they are changing orientation (such as C_d and effective frontal area), a multitude of trial-and-error inputs and calculations can be required to match the wreckage part locations found at the accident site with an estimated aircraft breakup altitude, direction, attitude, and speed condition.

FIGURE 6.5 Example of how wind affects pieces of wreckage with different ballistic coefficients.

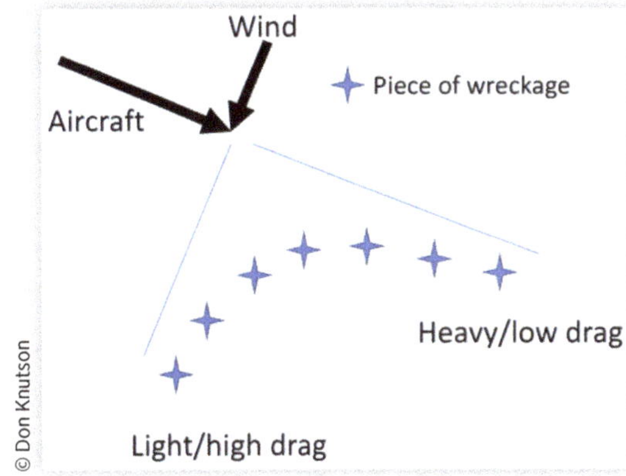

Some good old technical documents to reference are "Generalized Tables for the Calculation of Trajectory Curves for Bodies Moving in Air," Report LR-277 (1960), and "Trajectory Analysis in Aircraft Accident Investigation," Report LR-310 (1961), both from National Research Laboratories, Ottawa, Canada. These reports provide ways to calculate part trajectories resulting from in-flight breakups, including estimates of drag coefficients, wind drift, and terminal velocity.

Last, but not least in this section, pay attention to the relatively *intuitive* aspects of how parts of an aircraft will separate in-flight and scatter, because it may not be an actual in-flight breakup (as defined at the beginning of this chapter). For instance, refer to Figure 6.6 and consider the following related circumstances involving a helicopter crash:

- Clear skies prevailed, with winds from the south-southeast at 7 knots.
- Crash occurred at ~0430 local time (night flight) in remote and high desert terrain.
- Five witnesses reported seeing the helicopter at a distance prior to and during the crash – three of them specifically said they saw a search light.
- Several witnesses reported explosions and flames when the aircraft impacted terrain.
- Aircraft was equipped for GPS tracking – flight position was recorded about every minute, but not during the exact time of the crash.
- Only known elevated obstacle at the accident site was 55-foot high power lines, which were not severed.
- High rotor RPM during terrain impact was established.
- No evidence of bird strikes was found.

* Trajectory equations that describe ballistic motion are discussed in the article, "Trajectory Analysis for Accident Investigation," by Dennis A. Crider, NTSB, AIAA Modeling and Simulation Technologies Conference, 2015.

FIGURE 6.6 Simplified accident site and wreckage distribution diagram.

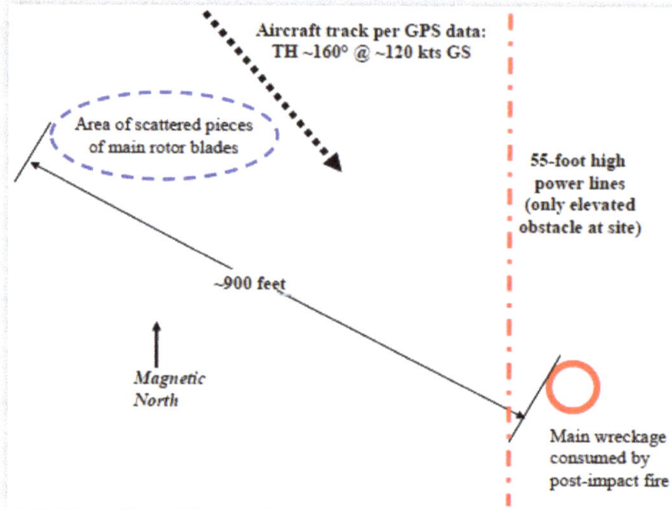

© Don Knutson

- Examination of the fragmented main rotor blades revealed a deteriorated leading-edge abrasion strip adhesive within the blade buildup.
- Review of maintenance records revealed that the leading-edge abrasion strip on a main rotor blade was replaced. Further investigation found that the mechanic used a propane torch instead of a heat gun to cure the bonding adhesive, which applied excessive heat that burned some of the adhesive.

The subject investigation report noted the cause of the crash to be from in-flight failure of the main rotor blades (for undetermined reasons). This theory would seem possible based on the main rotor leading-edge abrasion strip adhesive damage; however, the accident site features and wreckage scatter findings are not supportive. Imagine the main rotor blades *rotating* at ~348 RPM (within normal operating range, and maximum tangential tip speeds of ~725 ft/s) and then suddenly start to come apart during flight. Pieces of the main rotor blades would be slung radially outward and scattered on the ground in a random pattern, instead of being relatively concentrated in only one area and one side of the flight track. Based on the available investigation findings, the only feasible way the blade fragments would be scattered as depicted on Figure 6.6 is by the main rotor's plane of rotation intersecting an elevated stationary object, and systematically chopping a piece of the blade off during each half or full rotation throughout the fragmentation process. The only elevated obstacle in the area was the 55-foot high power lines. Trajectory analysis of fragmented pieces of the main rotor (based on their potential dimensions, weight, launch angle, and tangential speed) shows that they could have traveled up to 900 ft if they had separated when impacting the powerline in proximity to the main wreckage. In essence, the overall findings are not consistent with an in-flight breakup.

Piecing the Wreckage Together

Airworthiness standards per Federal Aviation Regulations (FAR) Part 23 for certified aircraft require a load factor limit of at least -1.52g to +3.8g for normal (non-aerobatic) category airplanes, and -3.0g to +6.0g for aerobatic category airplanes (negative g means feeling lighter such as when going over a hump on a roller coaster). Structural overload failures commonly occur at g-loads *exceeding* the aircraft's certified limits (sometimes by a factor of greater than 1.5), which means that engineering certification test results (e.g., for the wings and stabilizers) are needed to provide the actual positive and negative load limits (when the structure exceeds its elastic* range). Aircraft structures are normally subjected to tension, compression, torsion, shear, and bending (Figure 6.7). During an in-flight breakup, these forces are obviously magnified and typically reveal themselves in the airframe deformation and overload failures.

Lift generated by an airfoil is a function of velocity squared. This means the lifting force on a wing flying at 200 MPH is *four times* greater than flying at 100 MPH (with the same aircraft attitude and air density). We should also keep in mind that all airframe structures should be designed to somewhat elastically bend and twist within their normal operating range without creating any aerodynamic control or mechanical binding problems.

Distributed aerodynamic loads (or pressure distributions) should be considered when examining overall deformation or ultimate failures of aerodynamic structures (wings, horizontal stabilizers, stabilators, vertical stabilizers, canards, etc.). Figures 6.8 and 6.9 show a typical lift distribution along an airfoil cross-section and along the wingspan. Control surface (e.g., aileron or elevator) deflections will shift the lift distribution, which can affect the aeroelastic properties of the structure when the aircraft exceeds V_{ne}. For instance, deflections of ailerons or elevators during high-speed maneuvers can create twisting deformation on wings or stabilizers, respectively. Also, airplanes in a rolling and pitch-up recovery create an asymmetrical loading on the wings because the aileron deflected downward on the up-going wing increases the wing's camber.

FIGURE 6.7 Forces acting on aircraft structures.†

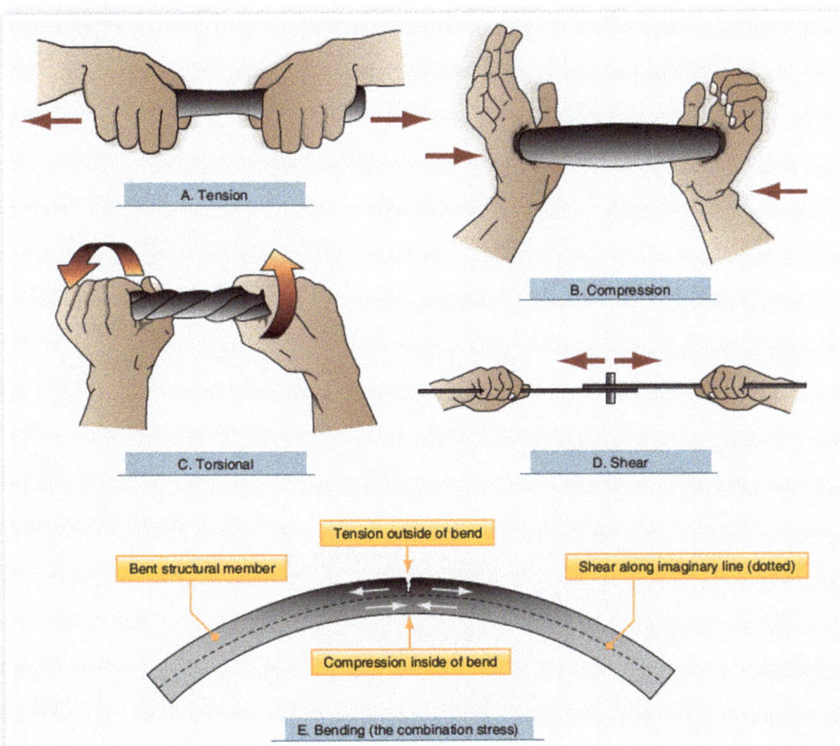

* Aircraft structures that maintain their elastic or aeroelastic properties will bend and/or twist and return to their original shape without sustaining damage.

† Depiction from FAA Aircraft Maintenance Technician Handbook – Airframe, Volume 1, Chapter 1.

FIGURE 6.8 Cross-section of typical subsonic airfoil showing pressure distribution along the percentage of its chord line (x/c). Note how the airfoil pressures increase with increased angle of attack (alpha).

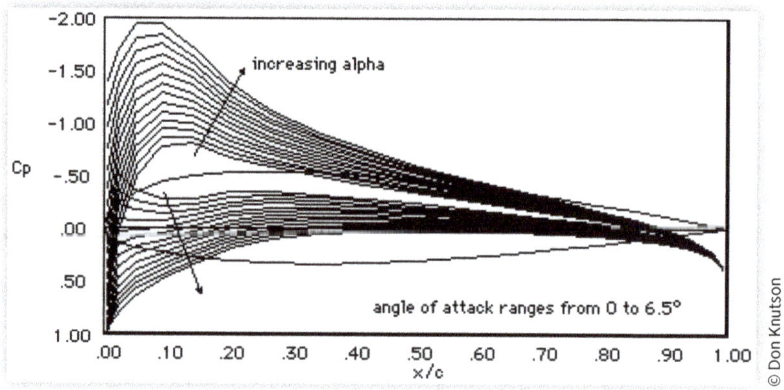

FIGURE 6.9 Approximate spanwise lift distribution on wings.

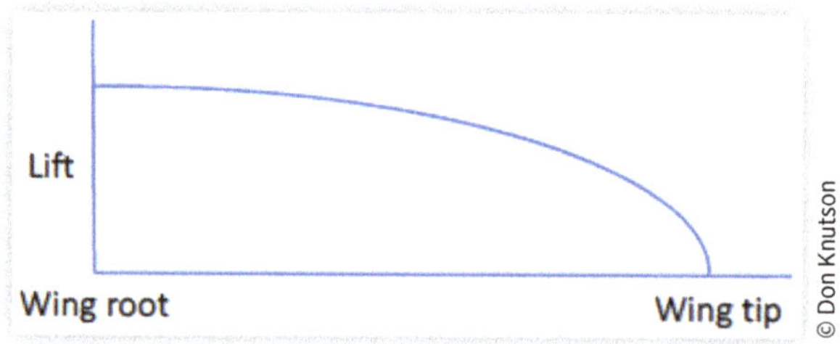

Pilots trying to maneuver aircraft while flying at the higher end of their capable operating range (e.g., maximum structural cruising speed [V_{no}]* or beyond velocity never exceed speed [V_{ne}]†) can ultimately lead to structural overload and in-flight breakups. This typically occurs when pilots try to regain control of the aircraft with abrupt or excessive control inputs during a high-speed dive or spiral. During a high-speed pull-up maneuver on an airplane, the wings are increasingly being aerodynamically loaded upward, which correspondingly creates an increased nose-down moment about the airplane's center of gravity. The tail section's pitch control surfaces (e.g., horizontal stabilizer and elevator, or stabilator) are aerodynamically designed to counter the nose-down moment by producing a tail-down force (Figure 6.10). Therefore, the wing separating in an upward or downward direction can tell us whether the wings or tail section separated first.

Negative/downward separation of the wings (an example is shown on Figure 6.11) indicates that the tail section (horizontal stabilizer and elevator, stabilator, or stabilizer and ruddervator) failed before the wings due to the initial loss of tail-down force causing the airplane to abruptly nose over. Evidence of this includes:

- Upper spar caps of wings exhibit tension loading/failure (refer to Figure 6.12).
- Lower spar caps of wings exhibit compression loading/failure.
- Positive loading of the wings in the plastic range can also exist; therefore, both upper and lower wing skins can exhibit buckling/wrinkling.

---

* V_{no} is the aircraft's airspeed that should not be exceeded except in smooth air and then only with caution (i.e., no abrupt or excessive flight control inputs that can cause a structural failure).

† V_{ne} is the aircraft's airspeed limit that may not be exceeded at any time.

FIGURE 6.10 Two-view diagram of an airplane showing aerodynamic lift forces during a pull-up maneuver (the airplane would abruptly pitch nose down without the downward lift force of tail).

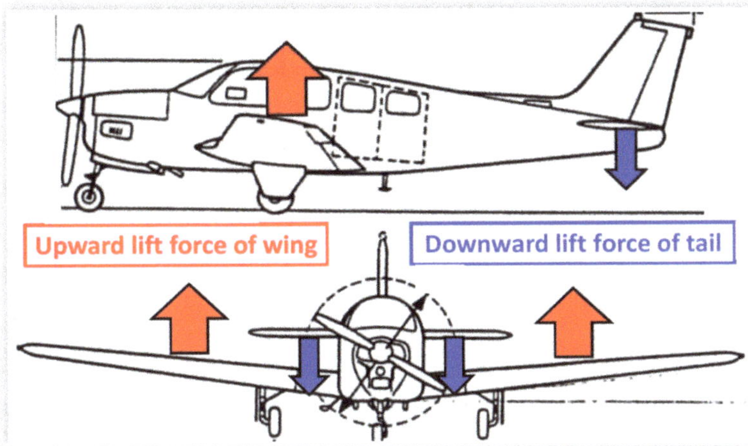

© Don Knutson

FIGURE 6.11 Airplane right wing that separated in a downward direction (wreckage lying inverted).*

© Australian Transport Safety Bureau

Positive/upward separation of the wings indicates that the wings failed before the tail section due to the tail-down force remaining in place at least until the wings separated. Evidence of this includes:

- Upper spar caps of wing exhibit compressive loading/failure.
- Lower spar caps of wing exhibit tension loading/failure.
- Upper skin on wings can exhibit buckling/wrinkling.

If just one wing separates, an extreme airplane roll rate can cause the aft fuselage and tail section to be twisted, which can result in the sides of aft fuselage skin exhibiting

* Australian Transport Safety Bureau (ATSB) Report AO-2017-029, In-flight breakup – Clonbinane, Victoria, Australia, July 31, 2007, VH-YJB, Rockwell International Aero Commander 500-S.

FIGURE 6.12 Fixed beam representative of a spar on a wing, horizontal stabilizer, or stabilator that is subjected to a downward bending load (such as shown in Figure 6.11), with the upper half loaded in tension and lower half loaded in compression (opposite circumstances when beam is subjected to an upward bending load).

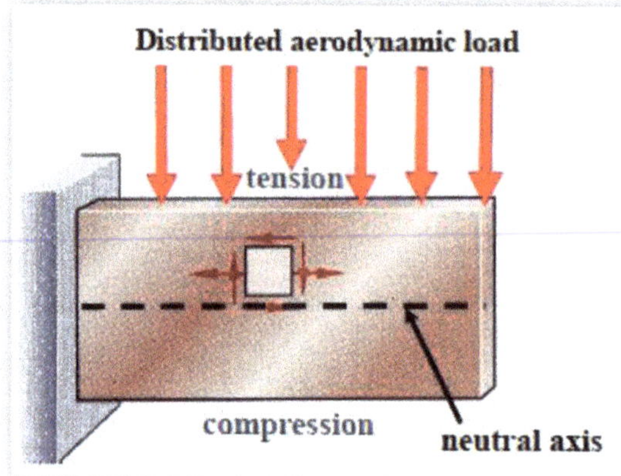

© Don Knutson

opposing diagonal buckling/wrinkling. Also, separation of the wings can result in its impact with the horizontal and/or vertical stabilizer leading edges (e.g., finding pronounced impact indentations and scrap marks).

Keep in mind that in-flight breakups caused from flying into extreme turbulence or a thunderstorm can result in structural failures that differ from the above-noted scenarios.

Aircraft diagrams such as shown in Figure 6.13 are also useful for documenting what parts separated during an in-flight breakup along with helping to evaluate the sequence of failure. This diagram involved an airplane wreckage inspection that revealed the

FIGURE 6.13 Airplane two-view diagram showing locations where parts separated during an in-flight breakup.

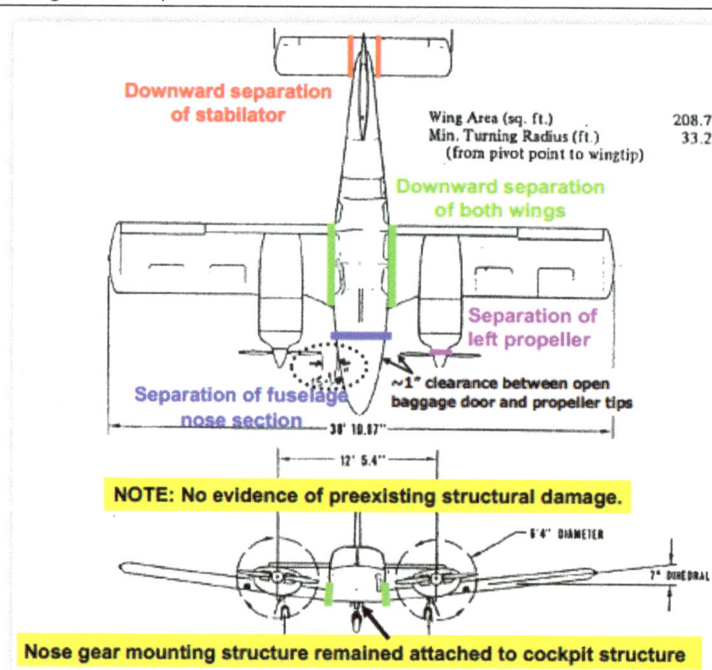

© Don Knutson

FIGURE 6.14 Parts of fuselage nose section, with its baggage door exhibiting at least two propeller slash marks.

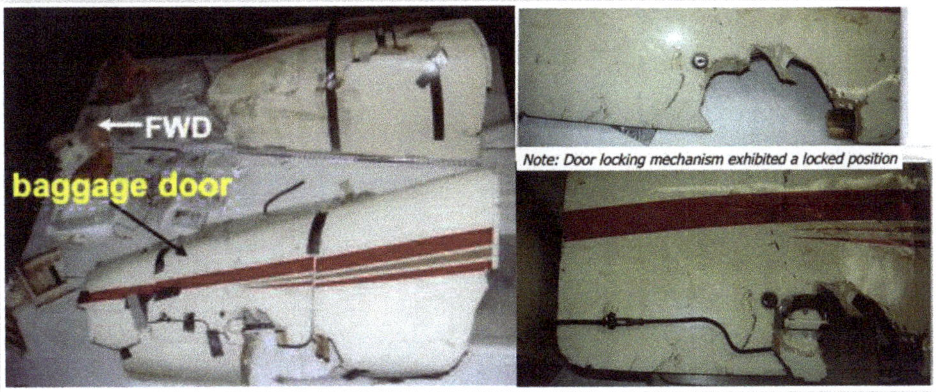

stabilator had failed downward before the wings failed downward and then the left propeller separated from its engine. The fuselage nose section baggage door exhibited two pronounced propeller slash marks from the left propeller blades (Figure 6.14). A left propeller blade tip impacted the nose wheel fork and piston assembly, and the blade's leading edge gouging matched the contour of the nose gear fork (Figure 6.15). The nose gear fork and piston separated from its nose landing gear cylinder (Figure 6.16) during the propeller blade impact.

The propeller slash marks found on baggage door from the left propeller blades prompted investigators to suspect that the baggage door came open during cruise flight causing the fuselage nose section to separate before the aforementioned stabilator/wings/propeller separation sequence. Is this a reasonable theory?

One of the ways we can answer this question is by considering that the nose landing gear cylinder and its mount structure remained fixed to the cockpit structure (i.e., not attached to the fuselage nose section) and then evaluating how the propeller's plane of rotation intersected/cut into the fuselage nose section. Using an approximately scaled front and top view diagram, we can match the left propeller's plane of rotation in relation to the fuselage nose section, the locations of the propeller slash marks on the baggage door, and the diameter of the left propeller – this results in an achievable intersection between the propeller tip and

FIGURE 6.15 Views of the one-foot sectioned left blade tip that impacted the nose wheel fork and piston assembly – the blade's leading edge gouging exactly matched the contour of the nose gear fork (please excuse the duct tape job).

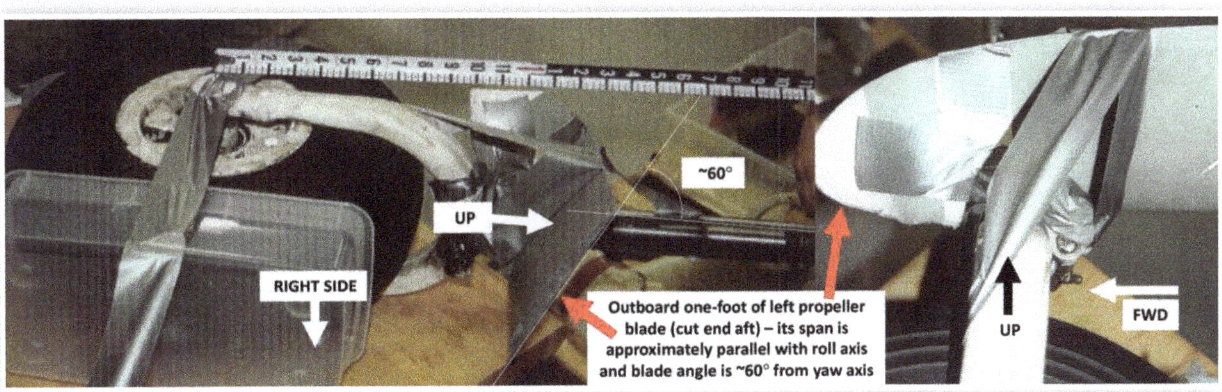

FIGURE 6.16 Side and front views of nose landing gear cylinder and its mount structure that remained attached to the cockpit structure.

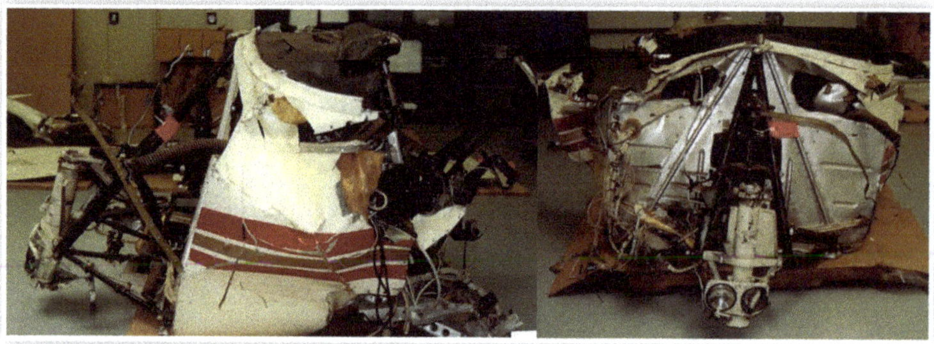

© Don Knutson

nose gear fork. Therefore, the fuselage nose section did *not* initially separate in-flight (contrary to the conclusion noted in the actual subject accident report that still exists in a database).

Flutter

Type-certificated* airplane designs are required to be free from any types of flutter when operating at designed diving speed (V_d)† in smooth air. However, modifications, repairs, or abnormal operating configurations that change the aerodynamic, mass, or stiffness characteristics of the aircraft can create a flutter condition within normal operating airspeeds.

Aeroelastic flutter is an unstable structural oscillation at a specific frequency where the airflow creates enough force to elastically torsion/bend the structure and self-perpetuate a cycled (e.g., up and down) movement. Once the oscillations/cycles start to weaken the structure or become divergent, the structure such as a wing or horizontal stabilizer can begin to permanently deform and/or come apart. Higher airspeeds (aerodynamic forces) usually increase the oscillation intensity. A relatively simple example of this is winds blowing across a flag – the stronger the wind, the faster and harder it flaps. Signs of oscillatory movements on fixed airframe structure that result in overload include X-shaped creases or "star" fractures in spar webs (e.g., diagonal buckling from upward bending, combined with opposed diagonal buckling in the same area from reversed/downward bending).

Control surface flutter can occur when the control surface becomes statically unbalanced due to external ice accumulation, internally undrained water that freezes, or after it is painted or repaired. Other causes include free/disconnected trim tabs (which occur within normal operating airspeeds) and turbulence-producing airframe configurations from deformed, separated, or protruding parts that are located upstream of the control surfaces. A good example of the latter is a left engine cowling door with worn-out latch springs on a Beech King Air 100 that came open and separated during cruise flight, which impinged on the adjacent wing leading edge and created enough turbulence to cause flutter and separation of the trailing outboard portion of the left elevator.‡

* Aircraft with a type certificate verifies that the aircraft is manufactured according to an approved design, which was in compliance with specific airworthiness requirements and standards (e.g., per 14 CFR Part 23 for airplanes with nine or less passengers and weigh 12,500 pounds or less and 14 CFR Part 25 for airplanes with 10 or more passengers and weigh over 12,500 pounds).

† V_d is defined in 14 CFR 23.335 and 25.335 – Design airspeeds, which is greater than V_{ne}.

‡ Transportation Safety Board of Canada (TSB) Aviation Report A95W0180, In-flight separation of cowling, Beech King Air 100, C-GNAA, 50 nm north of Edmonton, Alberta, on September 26, 1995.

Flutter on interconnected horizontal flight control surfaces like an elevator or stabilator that exhibit just their right or left side separated in a downward direction, and the opposing side fragmented in at least a couple pieces and showing both upward and downward deformations, typically indicates that flutter occurred *after* the first piece separated. Once one side of the control surface separates, the remaining hinged part of the control surface is no longer balanced and can violently oscillate/hammer on both control stops (Figure 6.17). This structural damage is sometimes misinterpreted as if the entire control surface had fluttered and caused the accident.

FIGURE 6.17 Stabilator fragmented in three main pieces and arranged together during a wreckage exam (top views), with its control stops exhibiting deformation and separation from being hammered (bottom view) after the left half of the stabilator separated downward in-flight (note downward deformation).[†]

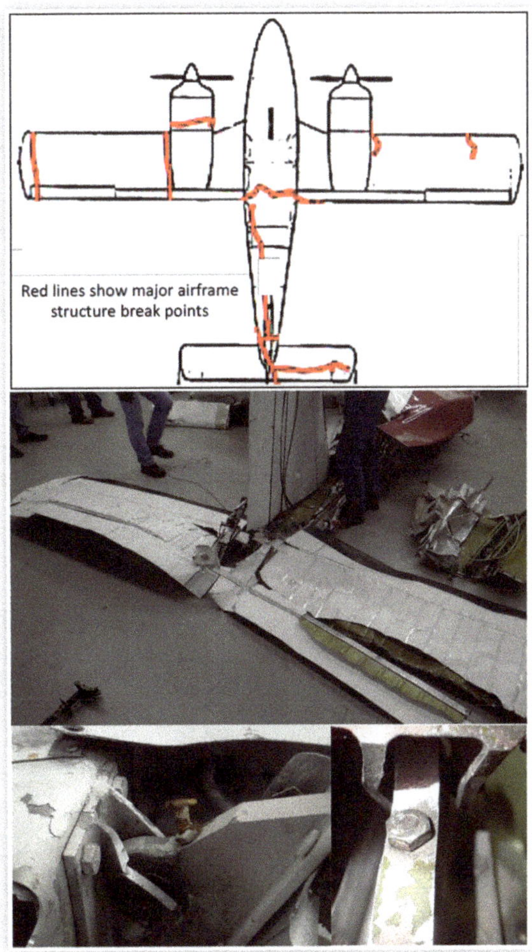

© Don Knutson

Whirl flutter is caused by the coupling of propeller or rotor aerodynamic forces with gyroscopic forces of the rotating mass (e.g., the propeller or rotor). This can occur on turboprop, tilt-prop-rotor, or rotorcraft aircraft structures. NASA research testing uncovered this problem after two Lockheed L-188 Electra commercial transport airplanes disintegrated during cruise flight.[†] To control whirl flutter, the basic concept is to maintain enough structural stiffness/strength to prevent the rotating masses from elastically displacing, and randomly oscillating from their normal rotational axis.

Aircraft Performance

Knowing the position of an aircraft along with the time it travels from point to point in three-dimensional space (latitude/longitude and altitude) allows us to estimate the speed, heading change, turn rate, climb or descent rate, g-load, bank angle, and pitch angle. For approximation purposes, we can assume a steady-state maneuver. These can be calculated by hand or using computer programs. Basic performance equations in the horizontal (e.g., X-Y or latitude/longitude) plane are as follows (in English units):

- $\mathbf{a_R} = \mathbf{V_t}^2/\mathbf{r}$, where $\mathbf{a_R}$ = radial acceleration in ft/s^2, $\mathbf{V_t}$ = true airspeed of aircraft in feet/second, and $\mathbf{r}$ = turn radius in feet
- $\tan \boldsymbol{\theta} = \mathbf{V}^2/(\mathbf{g_c} \times \mathbf{r})$, where $\boldsymbol{\theta}$ = aircraft bank angle in degrees and $\mathbf{g_c}$ = gravitational constant = 32.2 ft/s^2
- $\mathbf{F_{cen}} = (\mathbf{W} \times \mathbf{V}^2)/(\mathbf{g_c} \times \mathbf{r})$, where $\mathbf{F_{cen}}$ = centripetal force in pounds and $\mathbf{W}$ = weight of aircraft in pounds
- $\mathbf{G} = 1/\cos \boldsymbol{\theta}$, where $\mathbf{G}$ = g-load aircraft experiences during a coordinated turn (e.g., the wings of an airplane in a 60° bank angle support twice the airplane's weight)

* NTSB Docket ERA13FA253, Piper PA34-200T, N31743, accident near Johnstown, NY, on May 24, 2013.

† Braniff Airlines Flight 542 broke up in flight over Buffalo, Texas, on September 29, 1959; and Northwest Orient Airlines Flight 710 broke up in-flight near Cannelton, Indiana, on March 17, 1960. These accidents led to whirl flutter certification requirements.

FIGURE 6.18 Finding the radius of a turn using aircraft position data (red line is the flight path).

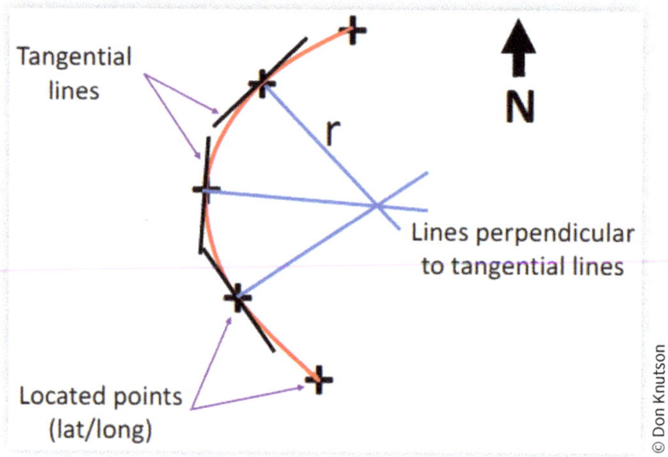

The radius of the turn can also be estimated by taking several points and curve-fitting* or "smoothing"† the flight path between them and then taking the tangents at each point to establish a perpendicular line. Where the perpendicular lines intersect reveals an approximate turn radius as well as a rate of turn based on the time interval between points (Figure 6.18).

Basic performance equations for aircraft performing steady-state pull-up maneuvers in a vertical (e.g., X-Z or Y-Z) plane are as follows:

- $\mathbf{q} = \mathbf{g_c}/\mathbf{V_t} \times (\mathbf{G} - 1)$, where $\mathbf{q}$ = pitch rate in radians/second
- $\mathbf{R} = \mathbf{V_t}/\mathbf{q}$, where $\mathbf{R}$ = pull-up radius in feet

Remember that a pull-up *with* a roll correction maneuver will asymmetrically load the wings.

Human Factors Causing Loss of Control

Flying at night without visual references on the ground or flight in instrument meteorological conditions with zero or limited visibility is the most common reason pilots become spatially disorientated.‡ Without being able to see anything outside the cockpit, the pilot needs to rely on flight instruments and know how to continually cross-check their indications, especially when they lose their attitude indicator. As you can see on

* When curve-fitting, a curvilinear path that exemplifies the subject aircraft in-flight is drawn that intersects and smoothly transitions between all the points (recorded aircraft positions).

† When smoothing a flight path, the recorded position of the aircraft is taken into account along with the previous two or more positions; then, the previous positions' heading, speed, and acceleration are used to predict/calculate for the next position.

‡ A good general reference is FAA Pilot Safety Brochure AM-400-03/1, Spatial Disorientation: Why You Shouldn't Fly By the Seat of Your Pants.

FIGURE 6.19 Flight test research report referenced on first footnote of this chapter.

GENERAL AVIATION PILOT PERFORMANCE FOLLOWING UNANNOUNCED IN-FLIGHT LOSS OF VACUUM SYSTEM AND ASSOCIATED INSTRUMENTS IN SIMULATED INSTRUMENT METEOROLOGICAL CONDITIONS

Pilot group (41 instrument-rated pilots)	Average time to recognize failure
Piper Archer A	*7.3 minutes*
Piper Archer B	*4.9 minutes*
Piper Archer C	*7.9 minutes*
Beech Bonanza A	*4.6 minutes*
Beech Bonanza B	*2.6 minutes*

Note:

- Archer A pilots used standard Archer instrumentation.
- Archer B pilots received 30 minutes of partial panel training on a computer simulator prior to their flight.
- Archer C pilots had a vacuum annunciator light as a warning when the vacuum system failed.
- Bonanza A pilots experienced an attitude indicator (AI) failure.
- Bonanza B pilots experienced both AI and horizontal situation indicator (HSI) failures.

Figure 6.19, the recognition time can be way too long before even instrument-rated pilots recognize an attitude indicator failure.

Long recognition times can cause the pilot to unknowingly follow a "dying" attitude indicator reading into an unusual attitude and subsequent loss of flight control. To avoid this, the pilot must first recognize that the attitude indicator is no longer reliable. The pilot must then interpret the indirect indications of attitude given by other flight instruments, such as changes in direction showing that the aircraft is in a banked attitude or increasing airspeed, which means that the aircraft is in a nose-down attitude.

Our vestibular system takes over when we no longer can see. Nature provided us with three semicircular canals in each inner ear that are oriented at right angles to each other so we can sense pitch, roll, and yaw. Each canal is filled with fluid with tiny upright hairs that sense movement of the fluid. The direction of the fluid/hair movement results in nerve impulses sent to the brain which interprets *angular acceleration* in each plane of motion. Moving our head in a "yes" motion affects the pitch sensation. Tilting our head from side to side affects the roll sensation. Moving our head in a "no" motion affects the yaw sensation.

Some key limitations of the semicircular canals are that they cannot sense a steady rotational motion below ~2°/s, and the hairs will return to an upright/neutral position in ~20+ s during sustained angular accelerations (i.e., the brain perceives that the acceleration has stopped). This is why vestibular illusions (without visual reference) include the "graveyard spiral," which begins with an insidious increasing bank angle that remains below the ~2°/s threshold. As the aircraft continues to pick up speed and increase its bank angle during the spiraling descent, the pilot will hear the rush of air/slipstream that

CHAPTER 6

FIGURE 6.20 Flight path showing where aircraft lost electrical power (in proximity to last recorded altitude) and then subsequently entered a spiraling turn prior to breaking up in-flight.*

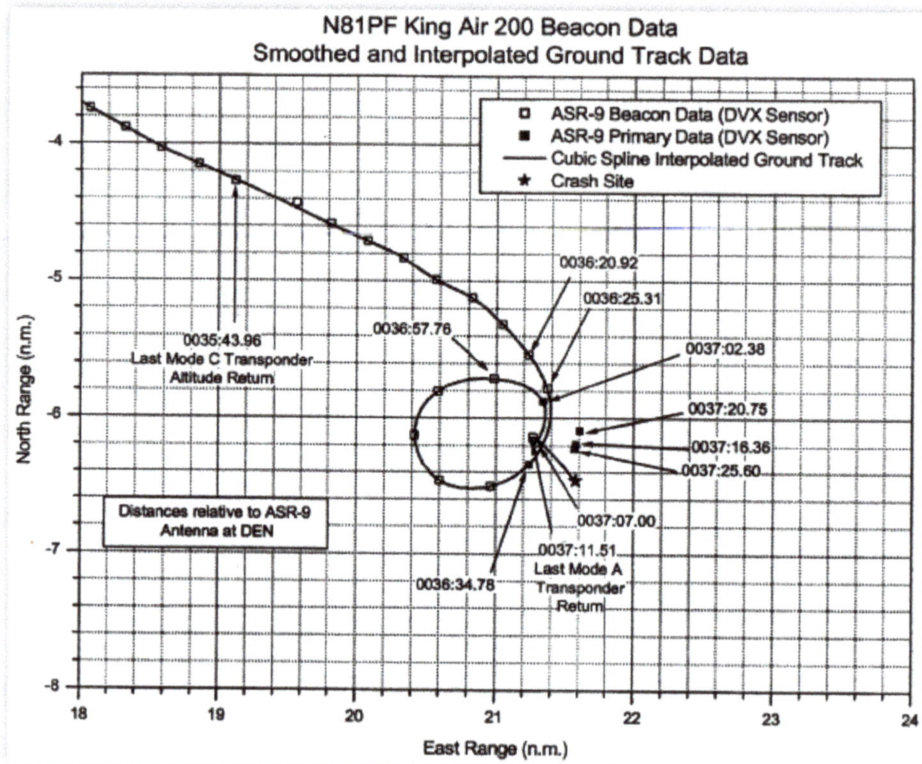

indicates the increasing airspeed of a dive and think the airplane is in a wings-level attitude. An attempt to pull out of the dive will only tighten the unrecognized turn and can overload the airframe structure. Figure 6.20 shows an example of a flight path tightening after the pilot experienced spatial disorientation.

Our vestibular system also has otolith organs that sense *linear acceleration* and consist of the saccule and utricle that have tiny hairs (similar to the semicircular canals) that sense direction of movement. The utricle senses change in horizontal movement, and the saccule senses vertical movement such as gravitational forces. These organs (again, without visual reference) can create the illusion of tumbling backwards or being inverted from a rapid transition from a climb to straight-and-level flight. Also, a pilot can mistakenly interpret forward acceleration during level flight as the nose being pitched up, which can prompt the pilot to unknowingly pitch the nose down and increase airspeed.

Another way our body senses angular and vertical movements is from the somatosensory system, which is made up of sensory receptors in the skin, muscles, and joints.

* NTSB/AAR-03/01, In-Flight Electrical System Failure and Loss of Control, Jet Express Services, Raytheon (Beechcraft) Super King Air 200, N81PF, Near Strasburg, Colorado, January 27, 2001.

FIGURE 6.21 Views of King Air B200 horizontal stabilizer and elevator that broke up in-flight and a shattered left windshield that started the rapid cabin depressurization.*

This can be related to the saying "flying by the seat of your pants," which also cannot provide an accurate orientation in three-dimensional space without visual reference.

Hypoxia, or oxygen starvation, occurs when pilots fly at altitudes where the atmospheric pressure is inadequate to force oxygen molecules into the bloodstream via the lungs. The main problem with hypoxia is pilots losing consciousness at high altitudes that allow the airplane to enter a high-speed descent/dive, which results in overload of the airframe structure when the pilots attempt a pull-up recovery. A good example of this occurred after an in-flight windshield failure at ~27,000 ft MSL, and the crew being unable to access oxygen and succumbing to hypoxia. Fortunately, the airplane descended to an altitude where they woke up and recovered the airplane, but not before losing a significant amount of the horizontal stabilizer and elevator (Figure 6.21).

---

* Photos obtained from NTSB Docket CHI07LA063, Raytheon (Beechcraft) Super King Air B200, N777AJ, near Cape Girardeau, MO, on February 2, 2007.

CHAPTER 7

Mid-Air Collisions

A few facts about mid-air collisions that would not necessarily be intuitive are that most of them occur in good weather/visibility conditions, and pilot recognition and reaction times (e.g., see-and-avoid procedures*) are typically underestimated. This chapter includes some important accident reconstruction aspects that blend into the wreckage reconstruction part of the investigation. Also, many of the investigative concepts presented in this chapter can be applied to aircraft ground collisions.

Pilot Awareness

Unless pilots are performing "kamikaze" air-to-air combat maneuvers, aircraft obviously fly into each other because the pilots do not know where each other are located. This is caused by a lack of awareness, which is usually the result of not seeing other aircraft, not communicating with other aircraft, and not having the proper instruments to detect other aircraft. Awareness is also a function of recognizing our physical and mental limitations while focusing on the tasks at hand. In this case, the more tasks, the more distraction from seeing and avoiding other aircraft. With all this in mind, investigators should try to obtain all the following information when a mid-air collision occurs:

- Flight track data from ATC facilities (e.g., radar, ADS-B, downloadable GPS, and primary flight display memory) – what direction, speed, altitude, climb rate, latitude/longitude, and airspace each aircraft is flying.
- Communication recordings and voice transcripts between controllers and involved aircraft; especially, look for periods of higher activity/workload during the accident

* Good references are FAA Advisory Circular (AC) 90-48D, "Pilots' Role in Collision Avoidance"; and FAA Safety Document P-8740-51, "How to Avoid a Mid Air Collision".

flight (e.g., more frequent exchange of radio transmissions or possible distractions being focused on inside the cockpit).

- Location of mid-air collision relative to other air traffic around the time of the accident - other pilots may have heard the accident pilots on the radio or saw the aircraft before, during, or after they collided.
- Distances of other airports closest to where the mid-air collision occurred – all airport approach and departure information should be obtained, including frequencies, elevations, and traffic patterns.
- Flight plans from the aircraft involved in the collision – departing airport, departure time, estimated time of arrival, altitude, and speed en route.
- Locations of mountains (or other terrain obstacles) that aircraft could have flown around/behind where radar coverage and radio communications (line of sight* to control towers, airports, or other aircraft) would have been affected at certain points along the aircraft's flight path.
- Aircraft flight operational policies if they are a scheduled charter or airline service – departure and approach procedure requirements.
- Onboard avionics (e.g., radio and navigation frequency settings, traffic collision avoidance system [TCAS], aircraft proximity sensor, and ADS-B).
- Aircraft windshield and window conditions (e.g., smearing, bug splatter, cracked, or crazed).
- Visual aids on the aircraft for alerting and spotting other air traffic (e.g., strobes flashing, position lights illuminated, and contrasting paint schemes).
- Pilots' visual limitations (e.g., prescription eyeglasses, color deficiencies, or use of sunglasses).
- Meteorological/sky conditions (e.g., visibility, cloud cover, and position of the sun or moon).

See-and-Avoid Issues

We need to understand the pilots' field of view and whether their viewpoints would have made any difference. Also, we need to be familiar with 14 CFR 91.113, right-of-way rules (except water operations) for aircraft, which are the following (verbatim in italics):

- *When weather conditions permit, regardless of whether an operation is conducted under instrument flight rules or visual flight rules, vigilance shall be maintained by each person operating an aircraft so as to see and avoid other aircraft. When a rule of this section gives another aircraft the right-of-way, the pilot shall give way to that aircraft and may not pass over, under, or ahead of it unless well clear.*
- *An aircraft in distress has the right-of-way over all other air traffic.*
- *When aircraft of the same category are converging at approximately the same altitude (except head-on, or nearly so), the aircraft to the other's right has the right-of-way. If the aircraft are of different categories –*
    - *A balloon has the right-of-way over any other category of aircraft.*
    - *A glider has the right-of-way over an airship, powered parachute, weight-shift-control aircraft, airplane, or rotorcraft.*
    - *An airship has the right-of-way over a powered parachute, weight-shifter control aircraft, airplane, or rotorcraft.*

* Radio waves normally used for radar and radio communications travel in a straight line and can be obstructed by terrain features.

- *However, an aircraft towing or refueling other aircraft has the right-of-way over all other engine-driven aircraft.*
- *When aircraft are approaching each other head-on, or nearly so, each pilot of each aircraft shall alter course to the right.*
- *Each aircraft that is being overtaken has the right-of-way and each pilot of an overtaking aircraft shall alter course to the right to pass well clear.*
- *Aircraft, while on final approach to land or while landing, have the right-of-way over other aircraft in flight or operating on the surface, except that they shall not take advantage of this rule to force an aircraft off the runway surface which has already landed and is attempting to make way for an aircraft on final approach. When two or more aircraft are approaching an airport for the purpose of landing, the aircraft at the lower altitude has the right-of-way, but it shall not take advantage of this rule to cut in front of another which is on final approach to land or to overtake that aircraft.*

The above-noted regulation contains reasonable rules to follow, but the words "vigilance" and "overtaken" need to be more clearly defined and evaluated when determining which pilot/crew should have been primarily responsible for seeing and avoiding the mid-air collision. In many cases, both pilots/crews could have helped in some way to prevent the accident. However, some valid questions to ask regarding the pilots' requirements to see and react to a pending collision are ... Was the field of view and scan time reasonable enough to detect the oncoming air traffic? Are the relative aircraft speeds too fast? Do aircraft with collision avoidance avionics (e.g., TCAS) have a greater responsibility than other aircraft that are not equipped?

When evaluating how to answer the above-noted questions, consider the following research data regarding recognition and reaction times (in seconds) for pilots confronted with a potential mid-air collision.*

Process/task	Response in seconds
See an object ahead	0.1
Recognize that object is an aircraft	1.0
Become aware that a collision course exists	5.0
Make decision to turn left or right	4.0
Muscular reaction	0.4
Aircraft lag in response to flight control input	2.0
Total time before aircraft begins to move	12.5

Keep in mind that the above-noted response times will vary (probably more toward the higher side) because of pilot distractions and visual limitations that commonly contribute to mid-air collisions.† Now let's consider the following three mid-air collision scenarios:

Example #1 involves a Cessna Caravan (10-seat, single-engine turboprop airplane, equipped with TCAS) transporting passengers to another airport, and a Piper Cub (two-seat, single-engine bush airplane) that was transporting a hunter to a remote camp – both were flying in Class E

* Obtained from FAA safety studies and flight test research performed by the US military.

† "See and Avoid/Cockpit Visibility," DOT/FAA/CT-TN89/18, October 1989 – this report summarizes notable facts about see and avoid and its relationship to cockpit visibility.

airspace* under visual flight rules (VFR) and in visual meteorological conditions. The Cub maintained its northerly heading and cruise altitude and was not required to be equipped with collision avoidance avionics nor establish air traffic control clearance or radio communication. Assume the closing speed between the two airplanes was estimated to be ~140 knots ground speed (~236 ft/s or 1 mile per 22.4 s). Should the Cub pilot be expected to see the Caravan that was traveling at twice the speed, while on a west-northwesterly heading and approaching from the right side, behind, and below the Cub? FAA flight safety training promotes a side-to-side scanning range of 120° (60° to the right and left of center) and 20° vertical scanning range (10° up and down from horizontal). Review of available GPS data shows that the Caravan pilot had the Cub within his 120° field of view leading all the way up to the mid-air collision (Figure 7.1), and he had enough time to recognize and maneuver away from the Cub. Also, the Caravan was outside the Cub pilot's 120° field of view from the time both aircraft were at least 1.2 nautical miles apart. Considering the overall circumstances, the Caravan pilot was overtaking and ultimately responsible to see and avoid the Cub.

Example #2 is a Lockheed F-16 and Cessna 150 operating within visual meteorological conditions.† Radar data showed the F-16 flying at 252 knots (ground speed) while starting a left turn toward the Cessna's flight path ~20 s prior to the collision and descending at ~150 ft/min. The Cessna was flying at 68 knots while maintaining heading and climbing 520 ft/min. Their closing speed, while on almost perpendicular headings just before impact, was ~255 knots (~430 ft/s or 1 mile per 12.3 s). In this case, the F-16 had a greater responsibility to see and avoid than the Cessna because [i] the F-16 clearly turned into the Cessna's flight path; [ii] Figure 7.2 shows that the F-16 looked like a small "dot" in the sky less than 8 s before the collision; and [iii] the Cessna pilot had less than 12.5 s to avoid the F-16 when it was 1 mile away.

FIGURE 7.1 Diagram showing Caravan pilot's 120° field of view relative to the Cub pilot's 120° field of view ~27 s before their mid-air collision.

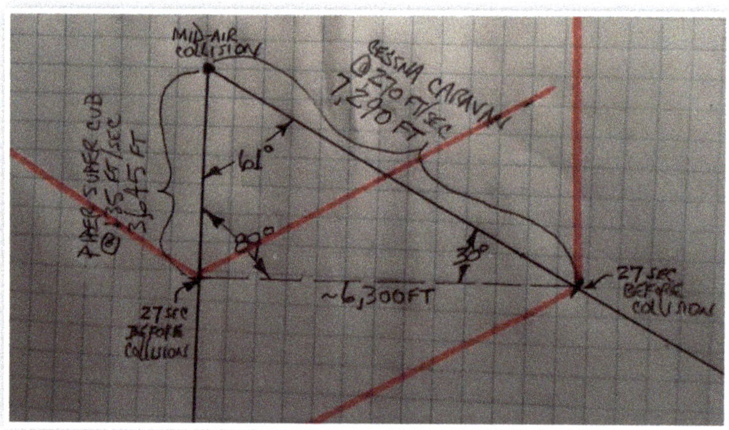

* Most airspace in the United States is class E, which extends from 1,200 ft above ground level (AGL) up to but not including 18,000 feet MSL (the lower limit of class A airspace). Areas where class E airspace begins at either the surface or 700 ft AGL are used to transition between the terminal and en route environments (around non-towered airports, which are designated on sectional charts). The airspace above FL600 (60,000 ft MSL) is also class E. No ATC clearance or radio communication is required for VFR flight in class E airspace. VFR visibility and cloud clearance requirements are the same as for classes C and D airspaces when below 10,000 ft MSL. Above 10,000 ft MSL, the visibility requirement is extended to 5 miles, and the cloud clearance requirement is extended to 1,000 ft below clouds, 1,000 feet above, and 1 mile laterally.

† Aircraft Performance Radar and Cockpit Visibility Study from NTSB Reports ERA15MA259A (C150) and ERA15MA259B (F-16), Moncks Corner, SC, July 7, 2015.

FIGURE 7.2 Pilot views of the F-16 from the Cessna 150 cockpit, with noted time before mid-air collision.

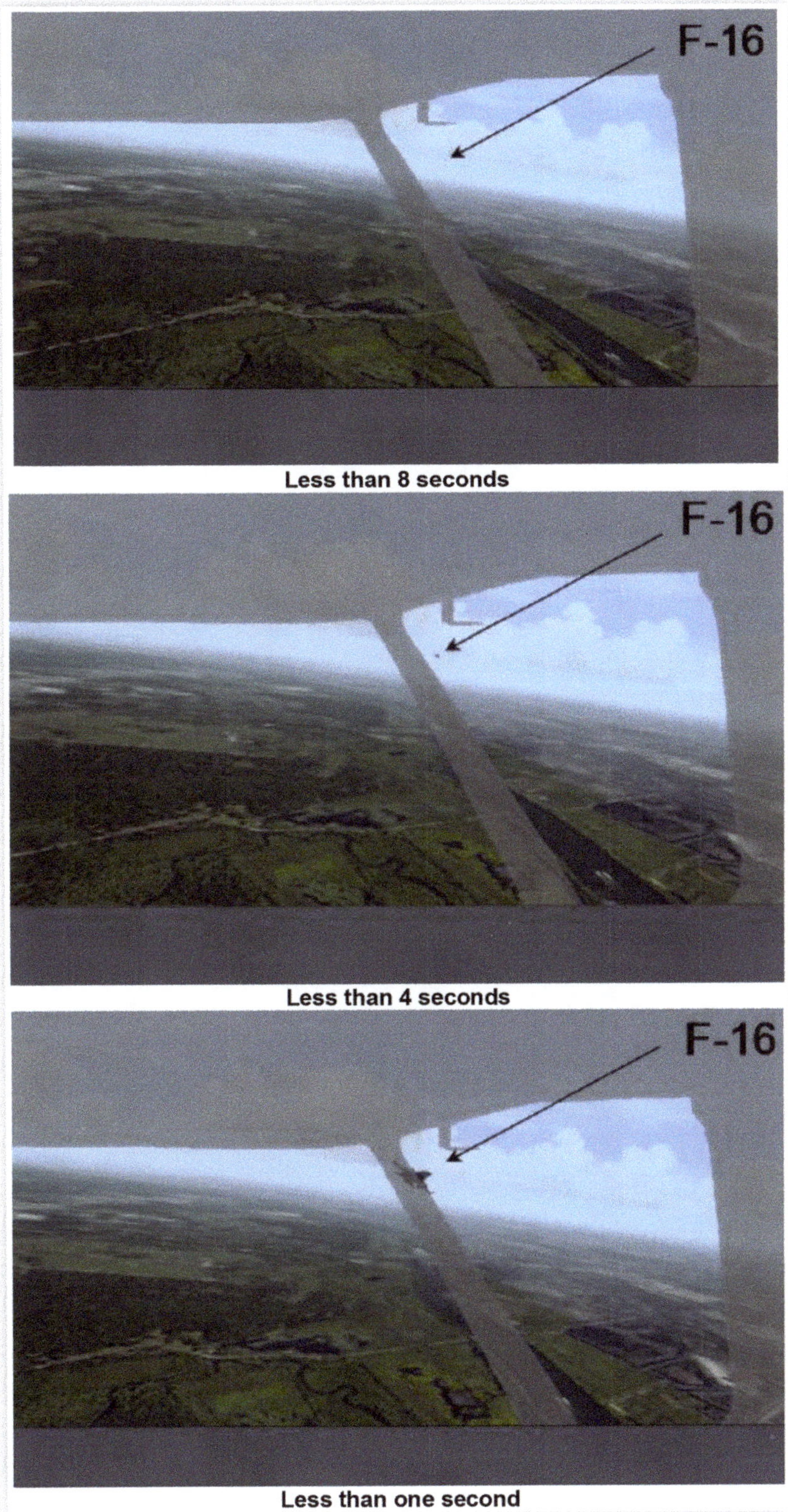

FIGURE 7.3 The pilots' viewpoint from the cockpit of a commercial jet airliner and small twin-engine airplane just seconds before colliding in mid-air.

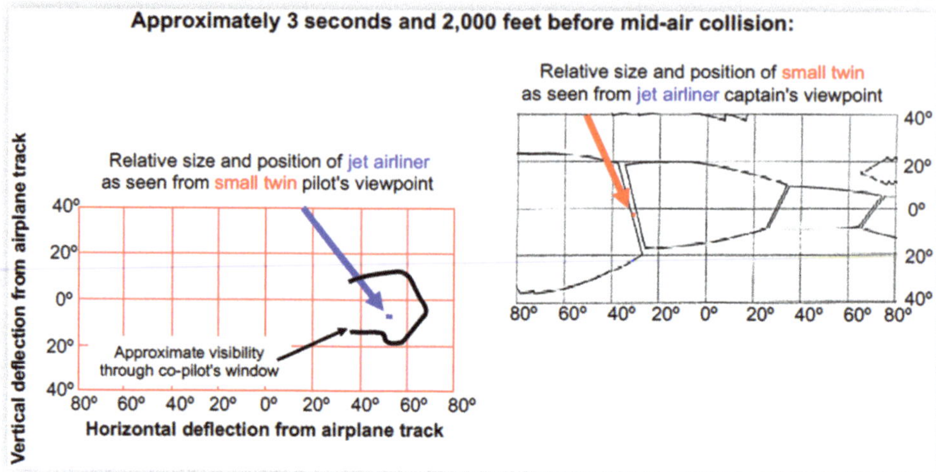

Example #3 involved a commercial jet airliner on a north-northwesterly heading at 329 knots (true airspeed) and a small twin-engine airplane on an east-northeasterly heading (approximately perpendicular to airliner's flight path) at 213 knots. This resulted in a closing speed of at least 392 knots (661 ft/s or 1 mile per 8 s). Figure 7.3 shows the pilots' viewpoints, which were similar to example #2 – the oncoming aircraft looks like a small dot to both pilots just seconds before impact.

In addition to the above examples, Figure 7.4 shows the relative size of an approaching airplane with a 10-m wing span.*

FIGURE 7.4 Pilot's visual/size perspective of an airplane with a 10-m wing span approaching at a closing speed of 200 knots and the time before a potential mid-air collision.

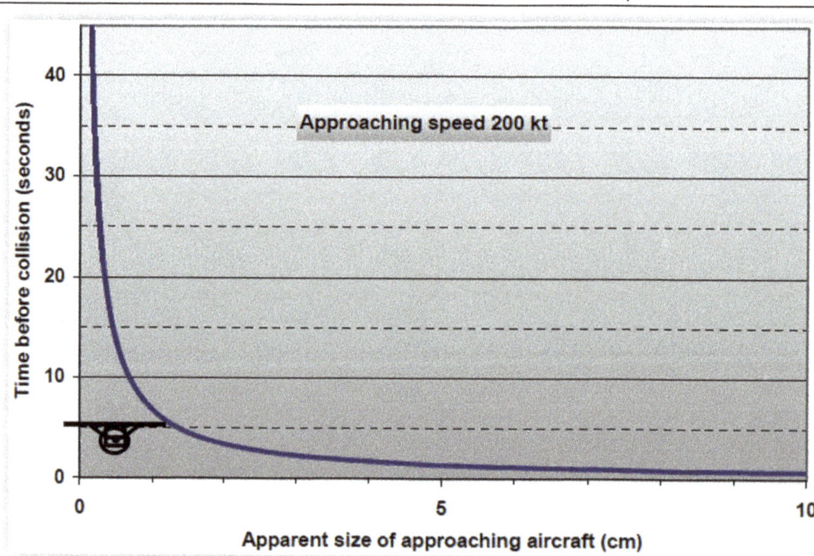

---

* Obtained from a safety study published and translated by the Bureau d'Enquêtes et d'Analyses pour la Sécurité de l'Aviation Civile (BEA, France) and involved 17 mid-air collisions that occurred over French territory from January 1, 1989, to June 30, 1999.

The above cockpit visibility studies show that aircraft flying at high closing speeds (i.e., that do not give the pilots enough time to reasonably recognize and maneuver away from oncoming air traffic) need collision avoidance avionics or ATC to alert them of a possible mid-air. With this in mind, unless the idea is to show how obviously difficult it is to see oncoming aircraft with inadequate time to change course, comprehensive cockpit visibility studies would probably be better utilized to evaluate aircraft flying at relatively slower closing speeds and/or in a more localized airspace (e.g., traffic pattern) where the pilots are typically more focused on anticipating aircraft in proximity to their flight path.

As Figure 7.1 shows, conducting a visibility study can be simply retracing the aircraft flight paths from the point of collision using position data and evaluating the general field of view from the converging aircraft. A comprehensive approach includes [i] documenting/modeling exemplar aircraft cockpits along with determining the pilots' seated position and eye level/viewpoint, [ii] determining aircraft position and movement in relation to each other in the airspace, and [iii] evaluating the pilots' visual limitations such as location of sun, atmospheric visibility, contrasting aircraft backgrounds, and color blindness.

Structural Impact and Scratch Marks

When evaluating two aircraft colliding in mid-air, we need to understand their relative movement (direction/heading, ground speed, and vertical descent or climb rate) prior to impact. Helpful information to estimate the collision angle includes the last few recorded latitude/longitude locations for each aircraft from GPS or radar data, and looking for the following damage signatures (if available):

- Orientation of the propeller or rotor plane-of-rotation cutting into the other aircraft
- Impact imprints matching the contact geometry of a part from the other aircraft
- Imbedded, trapped, or separated parts from the other aircraft
- Matching scratch, gouge, or paint transfer marks found on both airframes[‡]
- Rubber scuff marks from fixed/exposed landing gear tires

Note: Using clear sheet scaled overlays of front, side, and top views of both aircraft come in very handy to help visualize their orientation/engagement during the collision.

Let's assume a Piper Aerostar and Cessna 172 collided in mid-air while at their respective cruise speeds. During their wreckage inspections, scratch marks were found on the outboard lower surface of the left wing on the Piper, and the outboard upper surface of the right wing on the Cessna (Figure 7.5). Preparing a diagram as shown on Figure 7.6 helps us visualize and estimate the relative motion between the two airplanes. An estimated collision angle can be calculated via geometry of a triangle where the sum of all angles equals 180° and by using the law of sines. The closing speed between the two airplanes can also be estimated by using the law of sines. The *smaller scratch mark angle* found on the Piper compared to the Cessna indicates that the Piper was *flying faster* than the Cessna.

‡ Be cognizant of both aircraft's relative movement after their initial impact (which could result in misleading secondary scratch directions), and always look to support your findings with the other/above-listed ways to evaluate the collision angle.

CHAPTER 7

FIGURE 7.5 Mock-up views of scratch mark orientations found on wings (horizontally oriented surfaces) of a Piper Aerostar and Cessna 172 involved in a mid-air collision.

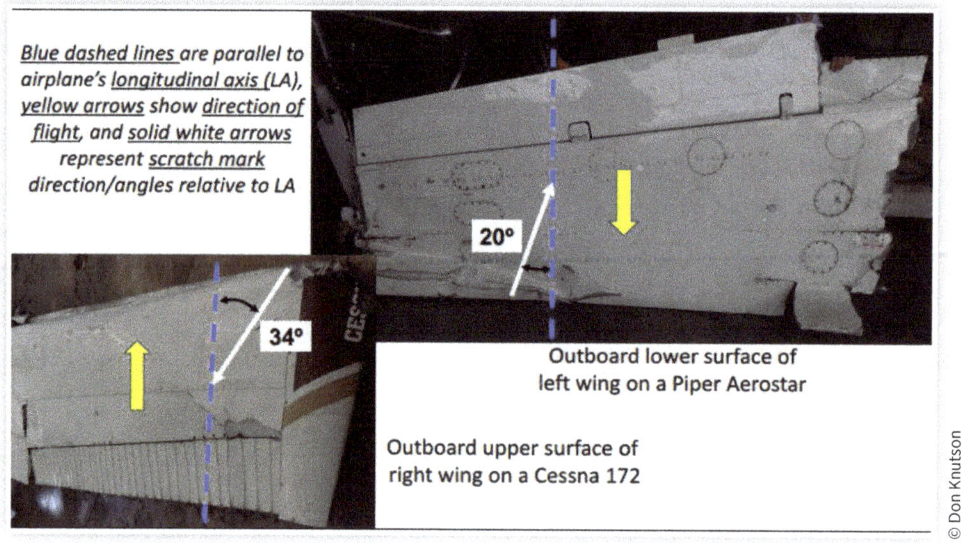

FIGURE 7.6 Diagram to show relationships of the scratch angles depicted on Figure 7.5 and the aircraft speeds, then using geometry and the law of sines to estimate collision angle and closure speed.

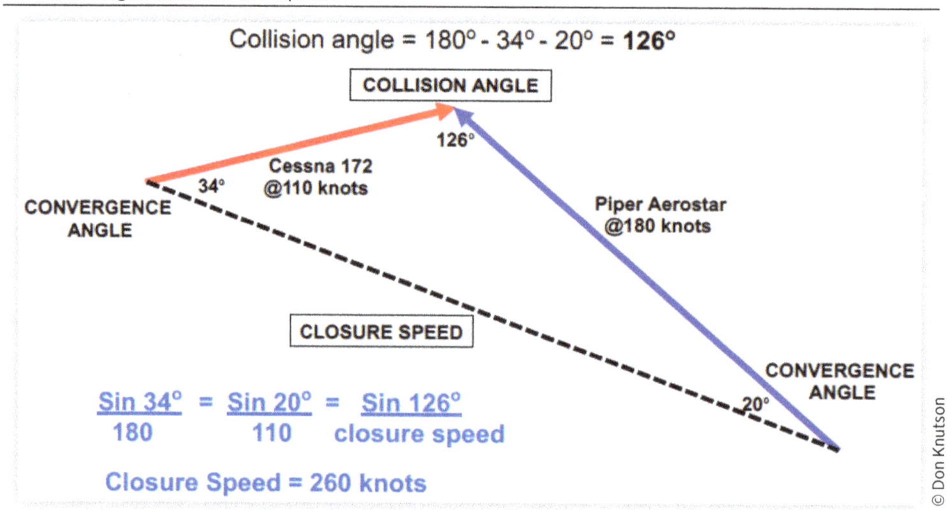

ICAO "Aircraft Accident and Incident Investigation", Part III, Chapter 10; and "Aircraft Accident Investigation," Chapter 17, by Wood and Sweginnis, Second Edition, ISBN: 9781892944177, also have helpful scratch mark orientation diagrams corresponding to various mid-air collision scenarios.

A good example of using flight track data along with evaluating propeller slash marks and matching impact imprints involved the previously discussed Cessna Caravan and Piper Cub mid-air collision (refer to Figures 7.7 through 7.11)*.

* NTSB Reports ANC16FA061A and ANC16FA061B.

FIGURE 7.7 Caravan and Cub on west-northwesterly and northerly headings, respectively, which indicates a collision angle of roughly 60°.

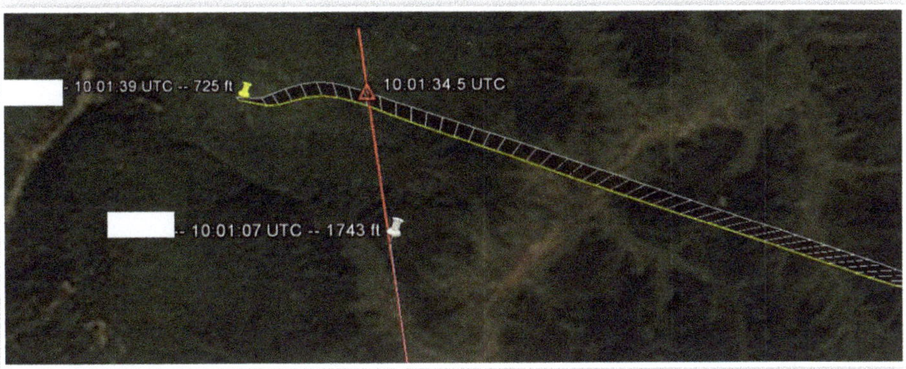

FIGURE 7.8 Caravan's left-wing station diagram with red line showing location of ~30° propeller cut, and thick blue line showing impact imprint with Cub's right-wing forward strut.

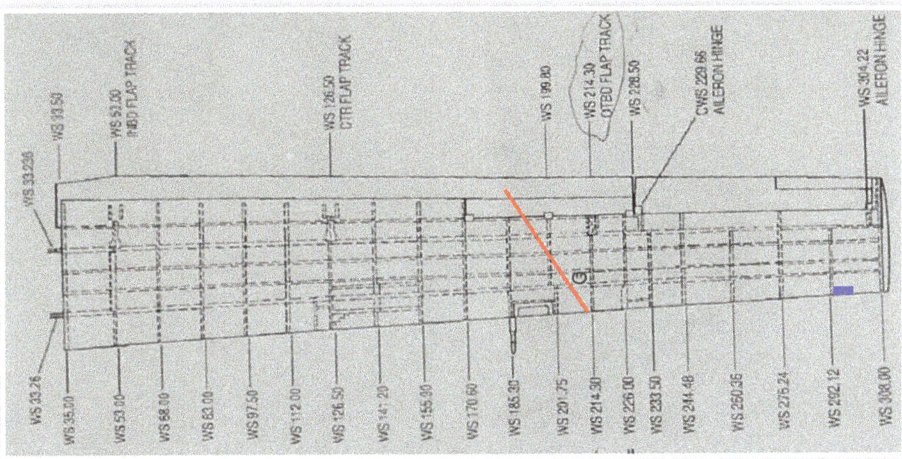

FIGURE 7.9 Subject Piper Cub involved in mid-air collision – red arrow shows where the Caravan's left-wing outboard leading edge impacted the Cub's right-wing forward strut (abeam to the cockpit area).

FIGURE 7.10 Views of where Caravan's left-wing leading edge (from WS 292 to 298) initially impacted the Cub's right-wing forward strut (imprint width and paint smearing on the leading edge matched).

© Don Knutson

FIGURE 7.11 Front, top, and left side views showing the orientation between the Caravan's left wing and the Cub's propeller (and its plane-of rotation) - red and blue arrows shows flight directions of the Caravan and Cub, respectively.

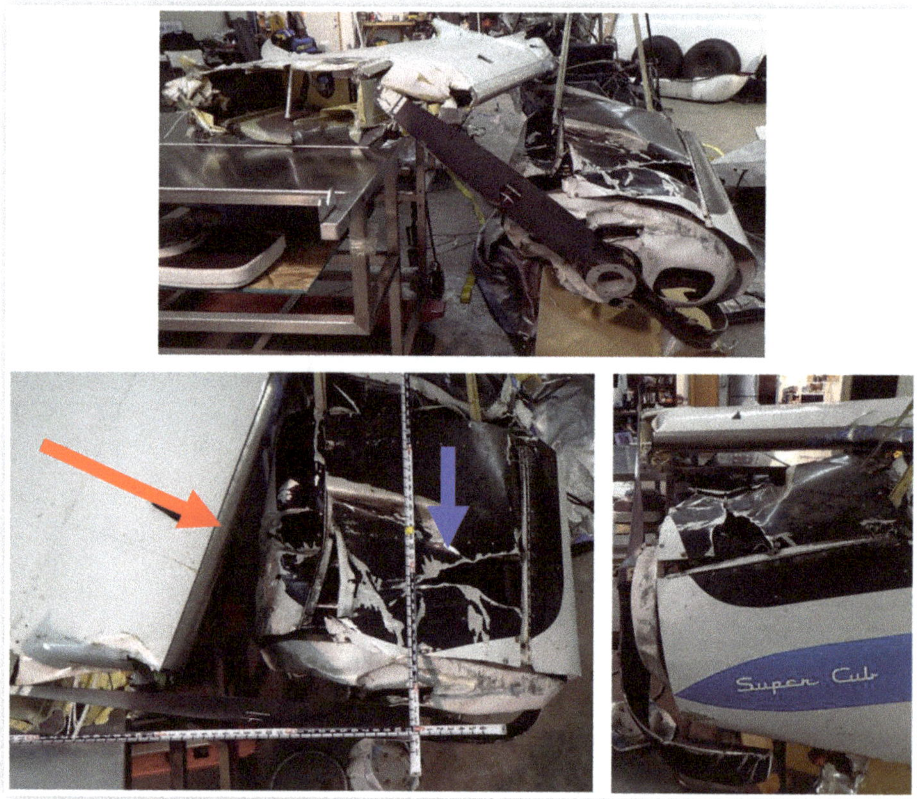

© Don Knutson

FIGURE 7.12 Scaled overlay drawings showing where the Caravan's left wing was slicing through the Cub's fuselage cabin when the Cub's propeller started cutting through the Caravan's left wing.

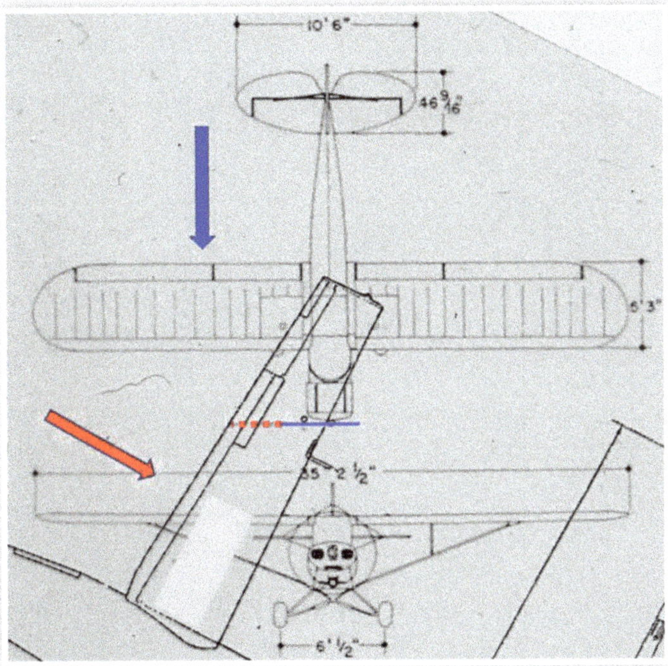

FIGURE 7.13 Vector diagram showing Caravan speeds relative to the Cub just before their mid-air collision.

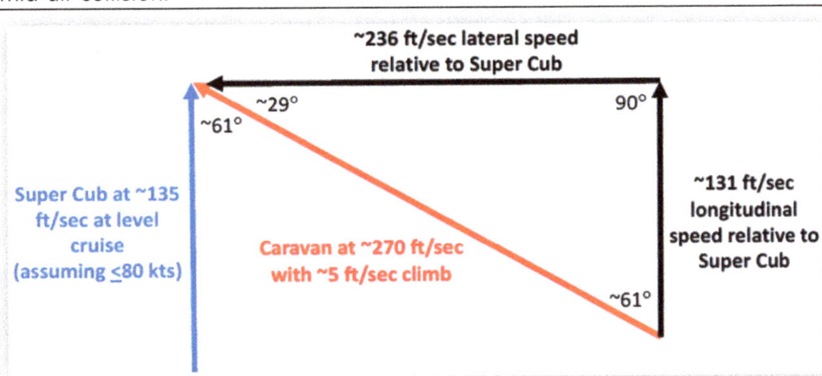

Figure 7.12 shows how to utilize clear sheet scaled overlay drawings of the Piper Cub and Cessna Caravan to visualize the sequence of events during their mid-air collision. In this case, the Caravan's left wing had already impacted the Cub's cockpit area before the plane of rotation of its propeller blades cut into the Caravan's left-wing structure. In addition to the overlay drawings, preparing a vector diagram that breaks down the relative speeds in the longitudinal/forward, lateral, and vertical directions (Figure 7.13) helps visualize/evaluate the relative motion between the two aircraft colliding.

CHAPTER 8

Fire and Explosions

This chapter discusses some basics of how to evaluate aircraft fire and explosion damage. Helpful publications that should be obtained for investigation references include the following:

- Aircraft Mishap Fire Pattern Investigations, Report #AFWAL-TR-85-2057, US Air Force Aero Propulsion Laboratory
- National Fire Protection Agency (NFPA) 921 - Guide for Fire and Explosion Investigations, ISBN 9781455916023
- NFPA 422M – Aircraft Fire and Explosion Investigators
- Forensic Investigation of Explosions (Second Edition), edited by Alexander Beveridge, ISBN 9781420087253
- FAA Advisory Circular (AC) 25.981-1C, Fuel Tank Ignition Source Prevention Guidelines

General Aspects of Aircraft Fires

The ingredients of a fire are [i] a flammable substance such as engine fuel or oil, hydraulic fluid, and baggage contents; [ii] an ignition or heat source such as an operating engine, engine exhaust and turbocharger failures, turbine compressor bleed air leaks, electrical sparks/arcs, thermal runaway of nickel-cadmium and lithium-ion batteries, and explosions; and [iii] air with an oxygen content of at least 12%* (air normally contains ~21% oxygen).

* DOT/FAA/ARTN-02/79, Limiting Oxygen Concentration Required to Inert Jet Fuel Vapors Existing at Reduced Fuel Tank Pressures, FAA Aircraft Research and Development, Fire Safety Branch, August 2003.

Some important characteristics/properties of aircraft materials we need to be aware of concerning fire include the following (which are amply documented and referenced in the above-noted publications):

- *Melting point*, which is the minimum temperature of a substance when it turns into a liquid/molten state.
- *Autoignition*, which is the minimum temperature of a substance that will initiate or cause self-sustained combustion (burning) independently of any sparks or other means of ignition.
- *Upper* and *lower flammability* or *explosive limits* of contained aircraft fuels, which is based on the percentage of concentration by volume of the fuel in air, within a temperature range (e.g., Jet A has a lower and upper limit of 0.6% and 4.9%, respectively, and a flammable mixture temperature range of 95°F to 165°F). These fuel/air mixtures will be too lean to burn below the lower limit and too rich to burn above the upper limit.

Aircraft fuel is the most common flammable substance that feeds post-impact fires due to breached wing tanks and fuel lines routed through the wings, fuselage, and engine compartment(s). Also, all aircraft fuels in an atomized/mist state are capable of being ignited at all ambient temperatures. This is not too surprising when considering FAA research has shown that the minimum spark ignition energy required to ignite hydrocarbon fuel vapor is 0.2 mJ (similar to static electricity that can be created by rubbing our stocking feet on carpet).

In many cases, fuel tank or line ruptures result in fuel spray that can readily ignite when contacting hot engine surfaces, arcs or sparks from damaged electrical wiring, and friction-generated sparks (between separated parts of the aircraft itself or parts of the aircraft scraping on terrain).

Electrical system problems such as arcing wires or overheated motors have been the primary source of cockpit and cabin smoke events. Also, aging aircraft research has shown significantly damaged and exposed electrical wiring problems* that can create fire and smoke hazards. Arced electrical wiring events are usually short-lived, but they produce temperatures well in excess of the melting points of copper, steel, and aluminum. Individual electrical wire breaks that exhibit “cup-and-cone” fractures and/or “necking” from being pulled apart are consistent with no current flow. Conversely, wire breaks with current flow can exhibit cooled globules (or little balls) of copper on the fractured ends and/or fusion with other wires – both created from arcing.

Keep in mind that smoke can end up being more of a threat than the fire itself to the crew or passengers due to inhalation and vision problems in-flight or while trying to egress the aircraft after the crash.

Post-Impact and Ground Fires

These fires are typically fed by fuel and oil that can reach temperatures between 1,600°F and 2,000°F, which is beyond the melting points of most aircraft materials and can easily

* National Institute of Aviation Research (NIAR) performed aging airframe teardown inspections on a 1969 Cessna 402A, with 19,699 hours (total time); 1979 Cessna 402C, with 25,547 hours; 1975 Piper Navajo Chieftain PA31-350, with 17,270 hours; and 1993 Beechcraft 1900D Airliner, with 15,203 hours.

FIGURE 8.1 Cessna 310 crash site showing the fuselage mostly consumed by fire.*

consume most the fuselage when crashing on land in a relatively remote area or where firefighters cannot readily access the crash scene (Figure 8.1). Strong winds (naturally blowing or accelerated from a venturi effect) and breached oxygen supply tanks and lines can elevate the flame temperatures similar to in-flight fires (as discussed below). Magnesium† is a combustible aircraft metal that can ignite during a post-impact fire, and create flame temperatures exceeding 5,000°F.

Maximum firefighting response times at an airport is required to be within 2–3 min when following 14 CFR Part 139, NFPA 403, Department of Defense Instruction (DODI) 6055.6, or ICAO Recommendation 9.2.19 and 9.2.20; therefore, aircraft should have the best chance of limited fire consumption when crashing on airports with designated fire stations. Aircraft crashing into water can also help preserve evidence, even when the aircraft is extensively fragmented during the impact. A good example of this was the Swissair 111 in-flight fire‡ that resulted in the airplane disintegrating when impacting the ocean at high speed, and about 98% of the aircraft was reportedly recovered. This limited fire consumption allowed investigators to reconstruct the wreckage on a three-dimensional frame, which led to finding detailed physical evidence of where the fire could have started and how it propagated above the interior ceiling of the forward cabin section and in the cockpit.

Titanium§ used in turbine engines can ignite at ~2,900°F (with ample airflow) from frictional contact between rotor and stator components. This can create engine fire temperatures exceeding 5,000°F, which can potentially burn through the engine casing and airframe structure in proximity to the engine installation.

In-Flight Fires

In-flight fires can reach temperatures over 3,000°F because of rapid airflow directed toward outside or inside parts of the aircraft, which feeds more oxygen to the fire. This can create a "cutting torch" effect that melts both steel and aluminum components

* Obtained from NTSB Docket CEN16FA172.

† Magnesium has an excellent strength-to-weight ratio (about two-thirds the density of aluminum).

‡ Transportation Safety Board of Canada, Report Number A98H0003, In-Flight Fire Leading to Collision with Water, Swissair Transport Limited, McDonnell Douglas MD-11, HB-IWF, Peggy's Cove, Nova Scotia 5 nm SW, September 2, 1998.

§ Titanium has an excellent resistance to corrosion and strength-to-weight ratio (about 40% lighter than steel).

FIGURE 8.2 Soot pattern trailing a cockpit smoke evacuation shutter on a Boeing 747* (left is forward).

contacted by the oxygen-enriched flames. Aircraft steel alloys (e.g., stainless steel) start to melt at ~2,550°F, which is beyond the typical ground fire temperatures; therefore, melted steel found at the accident scene is most likely evidence of an in-flight fire.

Black soot (from incomplete combustion of carbon compounds) attaches to downstream surfaces cooler than ~700°F and along the slipstream or direction of flight during an inflight fire (Figures 8.2 and 8.3). Some examples would be a cockpit/cabin fire with smoke exiting out the side of a fuselage opening; or a right engine fire on an airplane, with smoke billowing out its nacelle and depositing soot on the right wing and flap surfaces directly behind/downstream of the burning engine. More evidence of an inflight fire includes:

- Rapid airflow that separates and splatters molten metals such as steel and aluminum against cooler aircraft component surfaces downstream of the fire

FIGURE 8.3 Right wing nacelle exhibiting discoloration and sooting downstream of the engine fire.

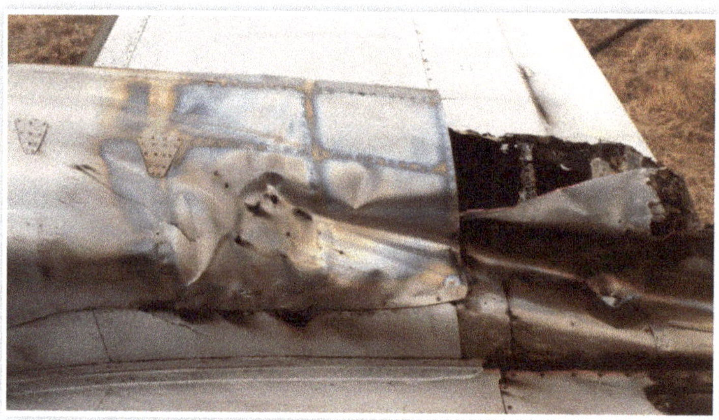

© National Transportation Safety Board

* Republic of (South) Korea Aircraft and Railway Accident Investigation Board, Aircraft Accident Report ARAIB/AAR1105, Crash Into The Sea After An In-Flight Fire, Asiana Airlines Boeing 747-400F, HL7604, International Waters 130 km West of Jeju Int'l Airport, on July 28, 2011.

FIGURE 8.4 Fractured ends of a wing spar structure showing "broomstraw."

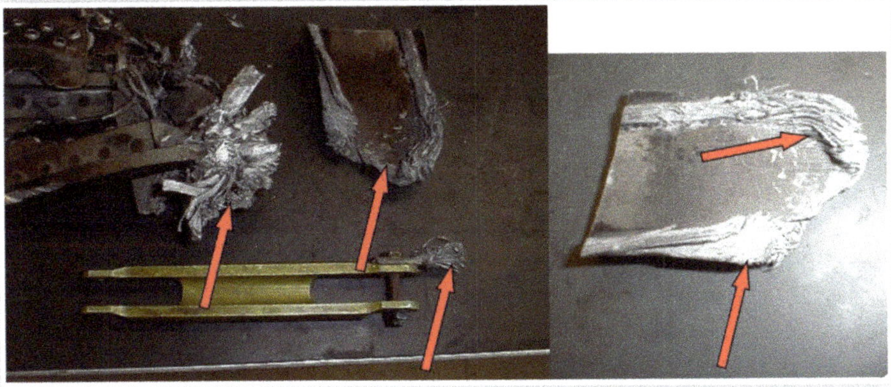

- Near-molten aluminum airframe structure subjected to shock loading or ground impact results in a "broomstraw" or "feather" appearance (Figure 8.4)
- Buckled/folded airframe wreckage such as the wing or fuselage skins that exhibit uniform soot or burn patterns both inside and outside of the deformation
- Matching soot patterns on normally exposed surfaces along with no soot on their fractured surfaces/edges (e.g., found on separated/torn airframe structure or pieces of broken windshield)
- Fire warning annunciator light bulb filament stretch

Determining Origin of Fires

In many cases, an aircraft fire results in such extensive material damage/consumption that its origin is not determined. Also, a fire's possible *points* of origin can easily become possible *areas* of origin depending on the amount of fire consumption. This usually presents a greater list of possibilities of what could have started the fire, which is not necessarily a disadvantage as long as safety improvements are made regarding all of the reasonable possibilities. For example, the origin of the in-flight fire on Air Canada Flight 797* was undetermined, but significant steps in air safety were taken to improve aircraft lavatory smoke detection, automatic fire extinguishing systems, emergency egress procedures, and firefighting capabilities.

Obtaining loading manifests and determining baggage contents (aerosols, ammunition, fertilizers, breathing oxygen, extra aircraft fuel, etc.) will help evaluate additional flammable materials and potential ignition sources onboard the aircraft.[†]

Aircraft wreckage not completely consumed by fire will exhibit discoloration, charring, blistering, distortion, and melting. Documenting the least to worst degrees of thermal damage (temperature and time that each part is exposed[‡]) will usually help map

* NTSB/AAR-86/02, DCA96MA070, In-flight Fire, Air Canada Flight 797, McDonnell Douglas DC-9-32, C-FTLU, at Greater Cincinnati International Airport, Covington, Kentucky, on June 2, 1983.

† Flammable liquids being shipped by aircraft need to comply with 49 CFR § 175.310 – Transportation of flammable liquid fuel; aircraft as only means of transportation.

‡ Controlled lab testing of exemplar materials is commonly necessary to verify specific thermal damage on aircraft components such as certain discoloration or rate of consumption over a finite time.

the progression/path of the fire. After the wreckage is documented at the scene, all the pieces should be gathered and laid out to properly organize and account for the recovered and consumed parts. In many cases, this requires a three-dimensional framework to more accurately locate and orient each piece of the aircraft structure. Then, the varying intensities of thermal damage can be further documented in sections or areas of aircraft structure and system assemblies. Structural/systems/component manuals and engineering drawings, along with an exemplar aircraft for making before- and after-fire damage comparisons will help with understanding the installations, operations, and material compositions. Some of the ways to characterize and evaluate the degree of fire damage to materials during the wreckage exam are as follows:

- Flames/heat and liquid vapors naturally rise when not being forced in any lateral direction by wind or other airflow.*
- Overall heat damage decreases as distance from the fire increases.
- Soot or clean burn† "V" patterns on relatively vertical/upward-oriented surfaces generally point toward the fire's location (Figure 8.5).
- Soot or clean burn "U" patterns on relatively vertical/upward-oriented surfaces indicate that the fire was below and farther away from the surface than "V" patterns.
- Inverted "V" soot patterns on relatively vertical/upward-oriented surfaces typically show where smoke was billowing out of a window or opening of an enclosed structure during a post-impact/ground fire.
- Amount of fire consumption (mass/loss of material) can give an indication of duration and intensity of the fire.
- Melting of materials transition from starting to soften, to fusion and viscosity (pieces combining and sticking together), to liquid flow, then finally vaporization.
- Steel and aluminum "B" nuts (common description of the nut that is used to connect metal tubing and flexible hoses) will loosen on their fittings when

FIGURE 8.5 Fire patterns showing the area of origin in proximity to and farther away from a surface.

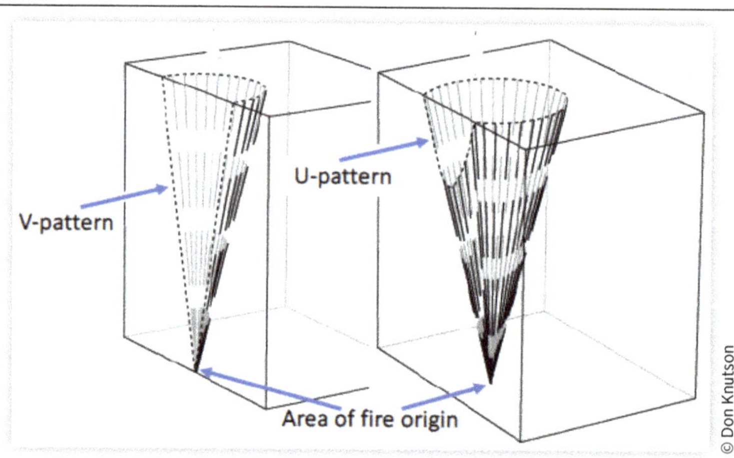

* In many cases, post-impact fires in aircraft fuselages oriented in a relatively level and upright position have resulted in most of the cockpit/cabin floor and subfloor structures remaining more intact than the sidewalls and ceiling. A good indicator of this characteristic is finding unconsumed paper materials such as manuals and logbooks located along the floor area.

† Paint, primer, and soot are burned away from the surface and look like "clean" metal that is bordered by soot.

subjected to excessive heat. The ones that survive a fire were probably torqued properly before the fire if they do not rotate to tighten more than one-quarter turn by hand (but this could require heat exposure testing on a mock-up installation).

- Pressurized fuel, engine oil, and hydraulic fluid lines exposed to 2,000°F have failed in less than 5 min without fluid flow (in the lines) and greater than 5 min with flow.
- Aircraft paint discolors at ~600°F, blisters at ~800–850°F, and burns off at ~900–950°F.
- Zinc chromate (green) paint primer turns to a tan color at ~450°F, tan to brown at ~500-600°F, and to black at ~700°F.
- Aluminum starts to soften and sag at ~800°F and melt at ~1,200°F.
- Stainless steel turns to a tan color at ~800°F and then progresses to light blue, to bright blue, and to black when reaching ~900°F.
- Copper (e.g., electrical wires) melts at ~1,980°F.
- Nylon and Teflon (electrical wire sheathing) autoignites at ~1,000–1,050°F.
- Titanium (usually associated with engine fires) turns light blue at ~900–1,000°F, to a dark blue at ~1,000–1,200°F, and then to a gray or yellowish color at ~1,200–1,500°F and melts at ~3,100°F.

Color-coding (with defined degrees of thermal damage) and vector diagrams can be used in addition to the wreckage layout to show the propagation of fire damage, which helps focus in on a possible origin of the fire. Recorded times from ATC communication transcripts when the pilot reported aircraft system failures, and any downloadable systems operational data during an in-flight fire can be compared to the determined path of the fire damage. The goal is to establish how fast the fire was propagating and an approximate timeline of events.

In many cases, specialists and experts in chemical, electrical, and material analysis are needed to help identify/verify/test flammable materials and oxidants and specific electrical problems, such as high-resistant faults, overloaded circuits, and bypassed system safeties (circuit breakers, fuses, etc.).

Types of Aircraft Explosions

The term "explosion" can be used loosely when describing other types of failures on aircraft such as a turbine engine overspeed causing rotor burst/fragmentation ("turbine wheel exploded"), or a rapid loss of fuselage cabin pressure ("explosive decompression"). For this chapter, we will define an explosion as the sudden conversion of potential energy (chemical or mechanical) into kinetic energy with the production and release of high-pressure gas[es], which then do mechanical work by moving, changing, or shattering nearby materials (per NFPA 921). Some other important definitions we need to be aware of include the following:

- *Deflagration* (e.g., from ignition of vapors inside a fuel tank) is the propagation of a combustion zone in an unreacted medium at a velocity that is *less than* the speed of sound (~1,116 ft/s or 340 m/s in standard day atmospheric conditions at sea level).
- *Detonation* (e.g., from a high explosive material in a bomb or missile) is the propagation of a combustion zone in an unreacted medium at a supersonic velocity (typically at least 5 times greater than the speed of sound).

Damage Caused by Detonation

This type of explosion typically exhibits the following wreckage damage signatures:

- Structures closest to the origin of detonation typically show the most fragmentation or shattering effect (creation of notably smaller fragments as depicted on Figure 8.6).
- Shattered material is propelled by the supersonic blast wave that "pepper" objects (or people) in its path, and creates small impact craters or pits in the objects that usually contain remnants of the shattered material, and sometimes traces of the unburnt explosive material.*
- Melting, erosion, and tightly rolled edges are found where blast fragments punctured metal accompanied by extremely hot detonation gases (known as "gas-washing").
- Punctured sheet metal has outwardly curled, sharp fractured edges deformed away from the venting gases (known as "petaling").

FIGURE 8.6 Skin fracture diagram showing the most concentrated fragmentation in proximity to the origin of detonation.†

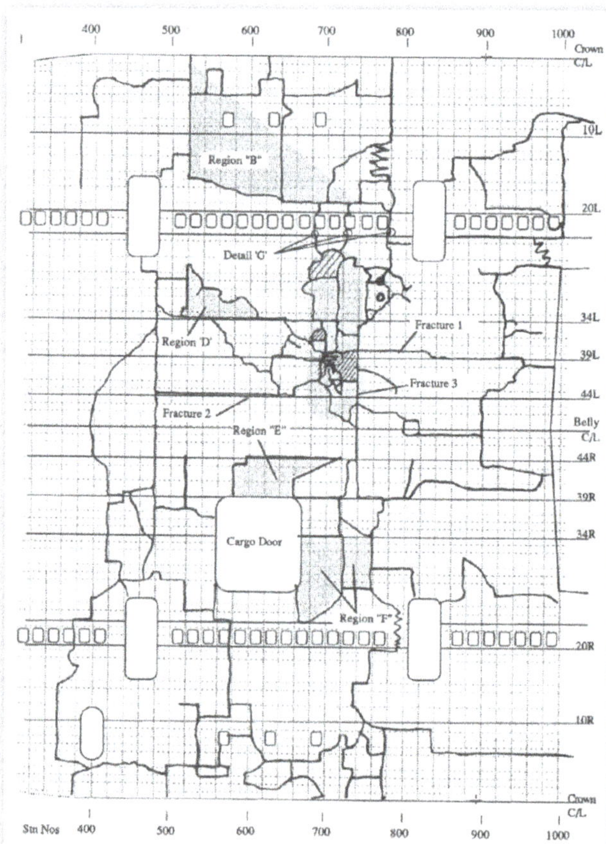

* Wreckage is tested for evidence of high explosives using chemical analysis and microscopic examinations, such as gas/liquid chromatography, mass spectrometry, and scanning electron microscope with energy-dispersive X-ray spectroscopy.

† Obtained from Air Accidents Investigation Branch (AAIB, United Kingdom) Report Number 2/90, terrorist bombing of Boeing 747-121, Pan Am Flight 103, N739PA, at Lockerbie, Dumfriesshire, Scotland, on December 21, 1988.

- Fractured edges of small metal fragments can exhibit "spikes" that are caused by the rapid/localized tearing and shearing of the material - they can be sharp and/or look wavy.

High explosive material residue and particles of unconsumed explosive material found on the aircraft wreckage can help focus in on the origin of detonation based on mapping the locations and amounts of material along with the direction it was propelled by the blast wave.

Damage Caused by Deflagration

Deflagrations commonly occur in fuel tanks or other structures where fuel has leaked into and contain fuel vapors that ignite within the fuel's upper and lower explosive limits (as defined earlier in this chapter). Specific causes of fuel tank system failures and sources of fuel vapor ignition are clearly outlined and referenced in FAA AC 25.981-1C. These explosion events typically exhibit the following wreckage damage signatures:

- Subsonic blasts result in more uniform deformation and distributed pressure, which create more "pushing" or "heaving" (versus shattering) effects on the exposed structural area (i.e., less material fragmentation occurs in proximity to the origin of deflagration than with detonation characteristics).
- Structure and objects in the path of the blast wave do not exhibit evidence of peppering, gas-washing, petaling, or spiking.
- Fuel-air explosions exhibit more widespread evidence of burning, scorching, and blistering.

When an aircraft encounters a fuel-tank explosion while on the ground, fragmented and relatively intact pieces of the aircraft found outside the main wreckage area can help determine the origin and cause of the explosion, providing effective firefighting response times significantly reduce the amount of post-impact fire consumption (as discussed earlier in the chapter). An example of this is a Boeing KC-135 air-refueling tanker that exploded on an airport ramp during routine maintenance (Figure 8.7)*.

FIGURE 8.7 Aerial view of KC-135 explosion and fire - notice how the damage is centered around the CWT (shown on Figure 8.9).

* Center wing tank explosion and fire on a Boeing KC-135 air-refueling tanker, operated by the Wisconsin Air National Guard, registration number 57-1470, at Milwaukee-General Mitchell Airport, on December 10, 1993.

FIGURE 8.8 Fragmented pieces of the KC-135 belly section (under the CWT) that were not consumed by the fire and found outside of the main wreckage area.

FIGURE 8.9 Diagram showing the general location of the KC-135 CWT.

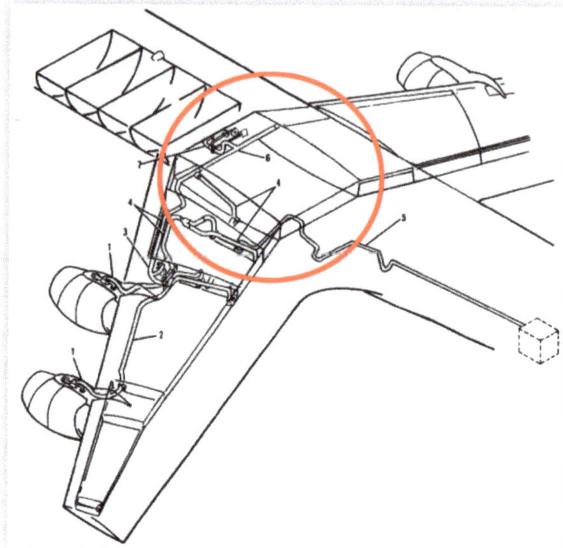

Fragmented pieces of the KC-135 belly section under the center wing fuel tank (CWT) were thrown radially outward and found outside of the main wreckage area (Figures 8.8 and 8.9). This helped determine the area of origin along with focusing in on the various CWT components as possible causes. Investigators determined the cause of the explosion was from nicked/exposed wiring in a fuel pump installed in the CWT that was arcing through the pump's casing while activated to transfer fuel – this ignited fuel vapors inside the tank that was close to empty.

conclusion

In addition to properly reconstructing wreckage, we want our reported investigative findings to be clearly understood and our analyses to stand up to the scrutiny of our peers and scientific acceptance. With this said, the following are some general considerations and advice when forming conclusions and preparing a report:

- This book provides basic methodologies and techniques on how to reconstruct aircraft wreckage *without* the benefit of onboard recorded flight or systems data. This furthers the potential need to substantiate investigative findings via custom ground or flight testing, as well as empirical knowledge derived from other related accident reports and past test and research data (e.g., systems and structural certification, full-scale crash results, human performance capabilities, and parts/materials subjected to specific conditions).
- Try to "connect all the dots" from the beginning to the end of an event by explaining how the various factual findings regarding the person-machine-environment interrelate (i.e., the anatomy of an accident). This does not include "cherry-picking" findings, which is using just selected facts to form a conclusion when one or more pertinent findings contradict the pet theory of what caused the crash or failure.
- Be careful when and how statistics are used to analyze accidents or failure trends. A humorous example of this is an experiment setup to determine the side effects of various mixed drinks. The first trial has a volunteer drink a bottle of bourbon with soda, which results in a hangover. The second trial has a volunteer drink a bottle of scotch with soda, which results in a hangover. The third trial has a volunteer drink a bottle of vodka with soda, which again results in a hangover. A potential theory: Soda contributes to hangovers! A more realistic example is selecting a certain aircraft make and model that has crashed more than any other aircraft within the same timeframe. Does this mean that this specific aircraft is the most dangerous to fly? It may, but before this conclusion can be made, we need to look deeper into the "why," such as establishing how the aircraft were being operated and maintained and whether the persons were properly trained to operate them in the given circumstances.
- In addition to referenced photos and diagrams enhancing the description of wreckage damage, our words should be comprehensive enough to provide the reader with a mental picture of what we examined. Also, words like "broken," "failed," "worn," and "burned" can be too ambiguous. Better ways to describe damage include fractured, elongated, gouged, buckled, and melted, *along with* its amount, location, and orientation. For example, noting that "The right wing exhibited deformation and leading edge damage" can be much better described as "The overall wing structure exhibited 10° upward deformation, with its leading edge, outboard of wing station 120, crushed aft 18 inches along its chord line." A technical report is definitely not the same as a fictional novel, but there is a lot to be said about how a best-selling author's words can immerse the reader into the scene of a story.

- We sometimes have to admit that the cause of an accident or failure is undetermined regardless of how comprehensively the situation is investigated. This is a tough pill to swallow when a ton of time and money has been spent trying to determine a finite answer. When this happens, investigators are sometimes compelled to "fit a square peg into a round hole" – to somehow justify all the work that was accomplished. Instead of force-fitting a conclusion, it is best to summarize the various reasonable possibilities that could have occurred based on the available evidence; then, maybe additional knowledge might come to light later on that can be used to better conclude what happened.
- Last, but not least, do not "shoot the arrow and paint the bull's-eye around wherever it lands" (as addressed in Chapter 1). This type of practice bastardizes the integrity of investigations.

In closing, I hope that this book will promote more aircraft crash investigators to pass on their experiences through presentations and publications that will be readily accessible to their peers. The frequency of crashes and failures involving aircraft occur much less than when compared to road vehicle accidents, and (as indicated earlier) each investigation typically presents its own unique set of circumstances. Also, some of our best teaching opportunities come from the mistakes we have made along the way. These are all good reasons why we should share our lessons learned to advance knowledge.

Blue skies!

appendix A

Aircraft Wreckage Reconstruction Terms

(in alphabetical order and not already defined in this book)

Abraded or **abrasion**: Scratched, scraped, or other frictional wear caused by materials being rubbed against each another.

Adhere or **adhesion** (in metallurgical terms): Melted metals sticking to another metal that has a cooler surface temperature (e.g., during in-flight fires or turbine engine failures).

Aerofoil or airfoil section: Cross-sectional shape of an object that moves through air and creates an aerodynamic force.

Airframe: The fuselage, booms, nacelles, cowlings, fairings, wings and other airfoil surfaces (including rotors, but excluding engine propellers), and landing gear of an aircraft, including their accessories and controls.

Airworthy: An aircraft or its component that conforms to its type certificate and is in a condition for safe operation. A non-type-certificated aircraft (e.g., experimental) is airworthy when it is in a condition for safe operation.

Alclad: Corrosion-resistant sheet aluminum comprised of high-purity aluminum surface layers (~5.5% thickness per side) metallurgically bonded and rolled onto a high-strength aluminum alloy core material.

Angle of attack: An acute angle at which a moving airfoil meets the airstream.

Appliance: Any instrument, mechanism, equipment, part, apparatus, appurtenance, or accessory, including communications equipment, which is used or intended to be used in operating or controlling an aircraft in flight, is installed in or attached to the aircraft and is not part of an airframe, engine, or propeller.

Autopsy: Examination of a fatally injured person to determine the cause and manner of death.

Biomechanics: The application of mechanical engineering principles to the human body, which includes evaluating the forces and motions during normal activities and injury-producing accidents.

Blistered: A coating, such as paint, that is separated from its surface and exhibits bubbling and/or rupturing due to being overheated, or parts of a metal surface that has separated/bubbled due to corrosive effects.

Brinelling: Permanent indentation of a hard surface due to an applied load (such as from ball bearings) being greater than the elastic limit of the bearing material (such as bearing raceways). It is named after the Brinell scale of hardness where a small ball is pushed against a hard surface at a preset level of force, and the resultant depth and diameter of the indentation indicates the Brinell hardness of the surface.

Brittle fracture: Fracture with little or no plastic deformation.

Buttock line (BL): A lateral distance (e.g., in inches or millimeters) to the left or right of the aircraft's longitudinal centerline (same principle as a wing station [WS] and horizontal stabilizer station [HS]).

Centripetal acceleration: A seemingly outward force caused by an object continually being pulled along a circular or curvilinear path.

Chord or **chordwise**: A measurement or location along the longitudinal distance between the leading edge and trailing edge of a wing, aileron, stabilizer, canard, elevator, stabilator, propeller/rotor blade, etc.

Coefficient of restitution (e): The ratio of the final to initial relative velocity between two objects after they collide, which normally ranges from zero (e.g., dropping a sack of sand on the ground) to close to 1 (e.g., billiard balls impacting each other).

Consumed: Burned and partially or fully melted.

Crack: A discontinuity in a solid body that has an initiation point (not always at the surface) and propagates to a finite size/length over time, depending on the orientation of loading and cycles applied. This propagation can continue until the body completely divides into two or more pieces.

Crash: Collision with terrain, obstacles, or other vehicles that typically results in an abrupt deceleration and impact-related damage.

Defect: Malfunction of a system or part that was not damaged as result of a calamity or improper use, and does not meet its intended use.

Detent: A catch or lever that initiates or prevents movement in a mechanism.

Deterioration: Decline in the quality of a device, mechanism, or structure over time due to environmental effects, corrosion, wear, or gradual changes in material properties.

Ductile fracture: Fracture with plastic deformation.

Elastic range: Deformation of a material under loading that can return to its original shape after the load is removed.

Embrittlement: Loss of ductility of a metal due to chemical or physical change.

Empennage: Also known as the tail section or tail assembly, which includes flight control surfaces (e.g., horizontal stabilizer and elevator, vertical stabilizer and rudder, stabilizers and ruddervators, and stabilator) that stabilize the flight dynamics about the lateral (pitch) and vertical (yaw) axes.

Experimental aircraft: Any aircraft that does not have a type design (e.g., home or kit-built) or is modified to where it does not conform to its type design and is in a condition for safe operation.

Fit: Ability of a part to physically interface with, connect to, or become an integral part of another part.

Flailing: Unrestrained/uncontrollable movement of parts of the human body during a crash or other accelerated condition, and typically starting from a seated position.

Flight safety-critical aircraft part: Any part of an aircraft containing a critical characteristic whose failure, malfunction, or absence could cause (i) a catastrophic failure resulting in loss or serious damage to the aircraft or (ii) an uncommanded engine shutdown resulting in an unsafe condition.

Forged or **forging**: Manufacturing process that uses hammering/compressive forces to shape metal typically in a heated state (and typically machined to further detail the part).

Form: Shape, size, dimensions, mass, weight, and other visual parameters that uniquely distinguish a part.

Fracture: Separation of material into two or more pieces when overloaded from tension, compression, bending, or twisting.

Fretting: Material wear created by tight-fitting surfaces allowed to rub against each other within a very small range of motion (e.g., from piston engine crankcase halves being under-torqued).

Fuel exhaustion: No fuel remaining in the tanks.

Fuel starvation: Interrupting fuel flow to the engine (e.g., fuel line blockage or improper fuel tank selection).

Function: Action that a part is designed to perform.

Fuselage: Main structure of an aircraft that contains the flight crew/cockpit, passengers/cabin, and cargo/baggage compartment[s].

Fuselage station (FS): Location along the centerline of the fuselage (e.g., in inches or millimeters) from a referenced datum/zero point that is set by the aircraft manufacturer at or near the nose of the aircraft.

Fusion: Melted metal that contacted and penetrated another metal surface because it was hotter than the surface's melting point (e.g., during turbine engine failures).

Galling: Metal surfaces sliding against each other with poor lubrication that creates overheating and metal transfer.

Gouging: Indentations or grooves in a material created during impact.

Hardness: Resistance to plastic deformation of a metal, and is quantified via indentation tests (e.g., Rockwell method for steels and Brinell method for castings and forgings) to determine whether it's properly heat treated.

Heat treated: A solid metal or alloy exposed to heating and cooling for a specific time to achieve softer or harder material properties.

Impact velocity: Speed and direction of an object during initial collision with terrain, an obstacle, or other aircraft.

Indicated airspeed: Airspeed indicator reading true speed through the air at standard sea-level pressure and temperature.

Injury mechanism: The manner or circumstance in which a physical injury occurred.

Joule: Unit of energy dissipated as heat when an electric current of one amp passes through a resistance of one ohm for one second.

Kinematics: A study of how objects move without reference to the forces that cause the motion.

Mach number: The ratio of true airspeed to the speed of sound.

Main rotor: The rotor that supplies the principal lift to a rotorcraft.

Major alteration: Changes to an aircraft that are defined and listed in Federal Aviation Regulations (FAR) Part 43, Appendix A.

Major repair: Repair to an aircraft that, if improperly done, can appreciably affect weight, balance, structural strength, performance, powerplant operation, flight characteristics, or other qualities affecting airworthiness (refer to FAR Part 43, Appendix A).

Materials science: The study of materials used in research, construction, and manufacturing; includes the fields of metallurgy, ceramics, plastics, rubber, and composites.

Metal cast or **casting**: A part fabricated by pouring liquid metal into a mold that contains a hollow shape or three-dimensional negative image of the intended shape, then cooled/solidified (this process is typically used for making complex shapes).

Modulus of elasticity: Ratio of stress (axial load divided by cross-sectional area) to the strain (length divided by change in length) within the material's elastic range in tension or compression.

Necking: Reduced cross-sectional area of a material from being stretched during tension overload (e.g., on electrical wiring and control cables).

Newton's first law of motion: An object at rest will stay at rest (in equilibrium) until an unbalanced force causes it to move or change direction.

Newton's second law of motion: The acceleration of an object is proportional to the net force applied to the object and inversely proportional to its mass (force = mass × acceleration).

Newton's third law of motion: When an object exerts a force on another object, the second object exerts a force equal and opposite to the force applied (e.g., the conservation of momentum).

Oil starvation: No oil available to lubricate the bearing surface, which leads to wiping and eventual disintegration of the bearing material from a prolonged engagement.

Overhauled: Restored to a serviceable or like new condition.

Pitch: Rotation about the lateral axis of an object (forward end or nose up is positive).

Pitot-static system: Involves the aircraft instruments that measure airspeed, Mach number, vertical speed, and altitude. This includes pitot tubes that receive ram air pressure, static ports that sense static pressure, air data computers (for glass cockpits), altitude encoders, cabin pressurization controllers, and the associated air tubing and electrical wiring.

Plastic deformation: Material deformed beyond its elastic range and cannot return to its original shape.

Polishing: Paint removed from a material during impact with terrain and creating a relatively smooth surface (e.g., propeller's leading edges).

Protruded: Extending beyond or above a surface.

Reconstruction: The science and art of determining causes and events from known circumstances and available physical evidence.

Repair: The elimination of damage or restoration of a damaged aircraft component that meets or exceeds its form, fit, and function.

Roll: Rotation about the longitudinal axis of an object (to the right is positive).

Scorched: Overheat damage to the object's exposed surfaces such as discoloration or blistering – this includes fractured areas and excludes melting of the object.

Scoring: Circumferential grooves in a bearing surface caused from foreign particles.

Sheared: Fracturing of a material by exerting forces parallel to the material's failure plane or cross-sectional area, or simply chopping or cutting through a material.

Span or **spanwise**: Measurement or location between the inboard end/root and outboard end/tip of a wing, aileron, stabilizer, canard, elevator, stabilator, propeller/rotor blade, etc.

Stall: When the angle of attack of a wing or airfoil is increased until there is a sudden loss of lift.

Stamped: A protruded object or surface that creates a corresponding imprint or indentation on another object or surface during impact.

Strain: The change in a material's length when subjected to a tensile or compressive load divided by its initial unloaded length.

Stress: Force exerted over an area, for example, pounds per square inch.

Stress raisers or **risers**: Discontinuities or changes in the contour of a material that increases/concentrates stress, and is where fatigue cracks start.

Supplemental type certificate (STC): A significant change to an FAA-approved aeronautical product from its original type design.

Terrain: What constitutes the ground surface (soil, rocks, fields, runways, roads, trees, etc.) and water surface (oceans, lakes, rivers, etc.).

Toxicology: Discipline overlapping with biology, chemistry, pharmacology, and medicine that involves the study of the adverse effects of chemical substances in the human body (e.g., screening and testing for drugs).

True airspeed: Speed of an aircraft along its flight path, with respect to the air mass (actual pressure altitude and air temperature) through which the aircraft is moving.

Type design: The engineering definition of a particular aviation product (an aircraft, engine, propeller, or appliance), which consists of its (i) drawings and specifications; (ii) dimensions, materials, and processes necessary to define structural strength; and (iii) limitations. FAA issues a type certificate after approving the type design.

Venturi effect: As airflow is forced through a smaller area, the air pressure decreases and correspondingly increases the velocity of the air.

Waterline (WL): Horizontal reference plane set by the aircraft manufacturer used to define vertical locations (e.g., in inches or millimeters) on the aircraft.

Windmilling: Unseized engine with an unfeathered propeller rotating solely via air being forced through the propeller's plane of rotation during flight.

Wiping: Displacement of bearing surface due to no lubrication and frictional overheating.

Witness mark: A surface scratch, indentation, smear, abrasion, or other marking on an object or part that can serve as evidence of its setting or position relative to other objects or parts.

Yaw: Rotation about the vertical axis of an object (forward end or nose to the right is positive).

appendix B

Handy Numbers/Conversions/ Equations

Temperatures:

- [°Celsius × 9/5] + 32 = °Fahrenheit (°F)
- [°F – 32] × 5/9 = °Celsius (°C)

Standard day atmospheric (air) conditions at sea level (SL):

- Density = 0.002377 slug/feet^3 = 1.225 kilograms/meter^3
- Pressure = 29.92 inches of mercury (Hg)
- Temperature = 59°F or 15°C

Changes in standard atmospheric conditions as altitude increases:

- Isothermal altitude = ~36,000 feet above SL
- Pressure lapse rate = –1.0 inch of Hg per 1,000 feet
- Temperature lapse rate = –3.57°F or –2.0°C per 1,000 feet (SL to isothermal altitude)

Distances:

- 1 statute mile = 5,280 feet = 0.869 nautical miles
- 1 nautical mile = 6,076 feet = 1.1508 statute miles
- Miles × 1.6093 = kilometers
- Kilometers × 0.6214 = miles
- Feet × 0.3048 = meters
- Meters × 3.2808 = feet

Pressures:

- Pounds/inch^2 × 2.036 = inches of Hg
- Pounds/inch^2 × 27.681 = inches of water (H_2O)
- Pounds/inch^2 × 0.068 = atmospheres
- Millibars or hectoPascals × 0.02953 = inches of Hg
- Pascals or Newtons/meter^2 × 0.000145 = pounds/inch^2

Speeds:

- MPH × 0.869 = knots
- Knots × 1.1508 = MPH
- MPH × 1.4667 = feet per second
- Knots × 1.6878 = feet per second
- Speed of sound ≈ 761 MPH or 661 knots true airspeed (standard day at SL)

Volumes:

- Gallons × 3.7854 − liters
- Liters × 0.2642 = gallons
- 1 cubic foot = 7.4805 gallons (US) = 28.3168 liters

Weights (standard day at SL):

- 100LL avgas ≈ 6.1 pounds/gallon
- Jet A ≈ 6.8 pounds/gallon
- Engine oil ≈ 7.4 pounds/gallon
- Water ≈ 8.21 pounds/gallon
- One cubic foot of air = 0.0765 pounds

Angles and some trigonometric functions:

- $1° = \pi/180 = 0.01745$ radians
- 1 radian = $180/\pi = 57.2958°$
- $\cos\theta°$ = adjacent side/hypotenuse (for right [90°] angles)
- $\sin\theta°$ = opposite side/hypotenuse
- $\tan\theta°$ = opposite side/adjacent side

Lift force: $L = ½\ \rho\ V_t^2\ A\ C_L$, where ρ = density of air (e.g., slugs/feet^3, and 1 slug = 1 pound-second^2/foot), V_t = true airspeed, A = planform area of wing or other lifting structure, and C_L = coefficient of lift of an airfoil

Modulus of elasticity (e.g., pounds/inch^2): $E = P\ L/(A\ \Delta L)$, where P is the axial load, L is the length of material, A = cross-sectional area, and ΔL = change in length during axial load

Kinematics for rectilinear motion:

- $V_1 = V_0 + a\ t$, where V_1 = velocity at end of t seconds, V_0 = initial velocity, and a = uniform or average acceleration (g_c = gravitational constant in vertical direction = 32.2 feet/second^2 or 9.81 meters/second^2)
- $d = V_0\ t + ½\ a\ t^2$, where d = distance traveled
- $d = (V_1^2 - V_0^2)/2\ a$
- $t = (-V_0 + [V_0^2 + 2\ a\ d]^{0.5})/a$

Kinematics for curvilinear motion:

- $\omega = \theta/t$, where ω = angular velocity in radians/second and θ = radians
- $\omega = 2\pi n$, where n = revolutions/second
- $V_{tan} = \omega r$, where V_{tan} = tangential velocity and r = radius
- $a_n = V_{tan}^2/r$, where a_n = normal acceleration toward the center of its path
- $\alpha = \omega/t$, where α = angular acceleration in radians/second^2
- $\omega_1 = \omega_0 + \alpha t$, where ω_1 is at end of t seconds and ω_0 is the initial angular velocity
- $\theta = \omega_0 t + ½ \alpha t^2$
- $\theta = (\omega_1^2 - \omega_0^2)/2\alpha$
- $t = (-\omega_0 + [\omega_0^2 + 2\alpha\theta]^{0.5})/\alpha$

Projectile motion (excluding air resistance):

- $t = V_0 t \sin\beta/g_c$, where t = time of flight and β = launch angle relative to horizontal in degrees
- $d_{ymax} = V_0^2 \sin^2\beta/(2 g_c)$, where d_{ymax} = maximum height reached
- $d_x = V_0^2 \sin^2(2\beta)/g_c$, where d_x = horizontal distance or range

Quadratic formula solving for x in $Ax^2 + Bx + C = 0$ (e.g., in triangular pulse [g-load vs. time] or above-noted kinematic equations): $x = [-B \pm (B^2 - 4AC)^{0.5}]/2A$

about the author

Don Knutson has performed hundreds of crash investigations domestically and internationally involving various models of civilian and military aircraft since 1991. He is President of Knutson Aviation Services (KAS, founded in 2002), a consulting firm specializing in aircraft (airplane and helicopter) accident and wreckage reconstruction, airframe and powerplant (piston and turbine) system failure analysis, and air safety education and research. Don has developed and taught college and continuing education courses on aircraft accident investigation (AAI) for the Wichita State University Aerospace Engineering (Wichita, KS) and Lewis University Aviation (Romeoville, IL) programs. For Boeing's military operations in Wichita, he instructed engineers in basic and advanced topics on supporting US Air Force crash investigations. Prior to starting KAS, Don was with The Boeing Company as a Flight Test Engineer where he evaluated fuel system designs and remote aerial refueling operations for the 767 Tanker program. He was the Director of Aviation for a forensics engineering firm where he provided AAI consulting and business development. Don also worked as an engineer and air safety investigator for Beech Aircraft Company where he evaluated mechanical system designs and was the engineering liaison between Beech and the NTSB, the FAA, and foreign government investigation agencies. He served as a Designated NTSB Representative for several international AAIs and helped train new air safety investigators during field investigations. Don is a full member of the International Society of Air Safety Investigators, a Technical Counselor and Young Eagles Pilot for the Experimental Aircraft Association, a certified fire and explosion investigator, and serves on a technical committee involving standards development of general aviation aircraft for the American Society for Testing and Materials International. He is an FAA-certified Airframe and Powerplant Mechanic (since 1981) and pilot. He received his Bachelor of Science in Mechanical Engineering from the University of Southwestern Louisiana (Lafayette, LA) in 1990.

index

www.ingramcontent.com/pod-product-compliance
Lightning Source LLC
LaVergne TN
LVHW070830220926
848486LV00030B/257